Programming the Raspberry Pi Pico in MicroPython

Harry Fairhead
&
Mike James

I/O Press
I Programmer Library

Harry Fairhead & Mike James,
Programming the Raspberry Pi Pico in MicroPython

ISBN:9781871962697

First Edition
First Printing, May 2021
Revision 0

Published by IO Press	www.iopress.info
In association with I Programmer	www.i-programmer.info
and with I o T Programmer	www.iot-programmer.com

For updates, errata, links to resources and the source code for the programs in this book, visit its dedicated page on the IO Press website: iopress.info.

Preface

The Pico is a remarkable microcontroller. It has a power and sophistication that would have been unthinkable just a short time ago. For the sort of jobs it is ideal for, it has plenty of processing power and enough memory to make tasks that would have once required careful planning, relatively easy. Instead of struggling with the machine, you can now focus on getting a good implementation of your algorithms.

The MicroPython language is a good choice to program the Pico. It isn't the fastest way, but in most cases it is fast enough to interface with the Pico's hardware and its big advantage is that it is easy to use. As a high-level language, MicroPython is based on Python 3, and is fully object-oriented. This means that you can create classes to encapsulate hardware and make it easier to use and understand. It is also easier to implement complex algorithms and so make your data processing easier. In general, you can take an existing Python 3 program and simply run it under MicroPython usually with no changes. If there are any changes then they are generally minor.

Another good thing about MicroPython on the Pico is that it is very easy to get started. A simple installation procedure and you have a working MicroPython machine which you can program almost at once using the Thonny IDE. Despite Thonny being a really useful IDE we also show how to use VS Code because it has more extensive syntax checking and input prompting. Both IDEs are easy to get started with and easy to use so the choice between them is up to you.

The Pico has so many resources that a comprehensive account would fill a book twice this size. In order to make things fit in the space available we have concentrated on things that are accessible from MicroPython and that are basic to getting things started. We have not covered "advanced" topics which generally lead the beginner into deep water far too quickly. The only major topic not covered is the use of the Pico's second core and threads. The reason is two-fold. The first is that this is still an underdeveloped area of MicroPython on the Pico and is likely to change. The second is that using threads entails a leap in difficulty and it is far too easy to create something that looks as if it works but contains a deep and significant problem. Put simply threading is dangerous and should not be introduced as a beginner's topic. What is more, in most cases you don't need it. If you adopt the approach of developing single-threaded programs until you are absolutely forced to consider introducing a second thread then your programs will be better for it.

We have, however, covered the use of the PIO (Programmable I/O) because it is one of the key advantages of using the Pico allowing you to delegate interaction with external hardware to a peripheral. Using this facility you can avoid any problems that you might have with MicroPython's lack of speed as you can write a PIO program which runs at the full speed of the Pico. The PIO isn't the solution to every problem, but it is very useful, challenging and a lot of fun. You will find out about the PIO in a dedicated chapter and then use it to implement a "driver" for the DHT22 and the 1-Wire bus.

If the Pico has a drawback, it is that it doesn't have a network connection. To solve this problem there is a chapter on using the low cost ESP8266 as a WiFi client and web server. The two devices together make the Pico a true IoT device.

This book doesn't teach you Python or MicroPython in the sense of basic programming, but a knowledge of how to program in almost any language is all you really need. All examples are written in a very simple style that avoids the use of some features of Python that are very "neat" but tend to obscure the meaning of the code from a beginner. You can easily refactor any of the examples into classes that suit your particular purpose and programming style.

This is not a projects book, although there isn't much left for you to do to round out the embryonic projects that are used as examples. Instead it is about understanding concepts and the acquisition of skills. The hope is that by the end of the book you will know how to tackle your own projects and get them safely to completion without wasting time in trial and error.

Thanks to our tireless editors Sue Gee and Kay Ewbank. Programming is the art of great precision, but English doesn't come with a built-in linter. Errors that remain, and we hope they are few, are ours.

For the source code for the programs in this book together with any updates or errata, links to resources including recommendations for obtaining electronic components, visit its dedicated page on the IO Press website: iopress.info.

You can also contact us at harry.fairhead@i-programmer.info or mike.james@i-programmer.info

Harry Fairhead
Mike James
May, 2021

Table of Contents

Preface 3

Chapter 1
The Raspberry Pi Pico – Before We Begin 11

The Pico And Its Family......12
What To Expect......14
What Do You Need?......14
Community......16
Summary......17

Chapter 2
Getting Started 19

Installing MicroPython......19
Hello World Using Thonny......21
Hello World VS Code......23
Summary......26

Chapter 3
Getting Started With The GPIO 27

Pico Pins......27
Basic GPIO Functions......29
Blinky......30
Summary......32

Chapter 4
Simple Output 33

Basic GPIO Functions......33
How Fast?......34
Including Pauses......36
Fixed Time Delay......37
Phased Pulses......38
Setting Multiple GPIO Lines......40
Summary......42

Chapter 5
Some Electronics **43**
Electrical Drive Characteristics....43
Driving An LED....44
LED BJT Drive....46
A BJT Example....47
MOSFET Driver....50
MOSFET LED....51
Setting Drive Type....52
Setting Output Mode....54
Basic Input Circuit - The Switch....54
Debounce....56
The Potential Divider....56
Summary....58

Chapter 6
Simple Input **59**
GPIO Input....59
Basic Input Functions....60
The Simple Button....60
Press Or Hold....62
How Fast Can We Measure?....63
The Finite State Machine....66
FSM Button....68
FSM Hold Button....69
FSM Ring Counter....70
Summary....73

Chapter 7
Advanced Input – Events and Interrupts **75**
Events....75
Interrupts Considered Harmful?....76
Hardware Events....77
An Edgy Button....79
Measuring Pulses With Events....81
Interrupts....82
How Fast Is An Interrupt?....84
Race Conditions and Starvation....86
Responding To Input....87
Summary....88

Chapter 8
Pulse Width Modulation **89**
Some Basic Pico PWM Facts....89
Pico PWM....91
Changing The Duty Cycle....95
Duty Cycle Resolution....96
Uses Of PWM – Digital To Analog....97
Frequency Modulation....99
Controlling An LED....100
What Else Can You Use PWM For?....102
Summary....104

Chapter 9
Controlling Motors And Servos **105**
DC Motor....105
Brushed Motors....106
Unidirectional Brushed Motor....108
Unidirectional PWM Motor Controller....110
Bidirectional Brushed Motor....112
Bidirectional Motor Software....116
Using A Single Full H-Bridge As Two Half H-Bridges....117
Controlling a Servo....118
Brushless DC Motors....121
Stepper Motors....124
Stepper Motor Driver....127
Stepper Motor Rotation – Using Timers....131
Summary....136

Chapter 10
Getting Started With The SPI Bus **137**
SPI Bus Basics....137
Pico SPI Interfaces....140
The SPI Functions....140
Using the Data Transfer Functions....142
A Loopback Example....143
Using The CS Line....145
The BME280 Humidity, Pressure and Temperature Sensor....146
Problems....151
Summary....152

Chapter 11
A-To-D and The SPI Bus 153
Pico ADC....153
The MCP3008 SPI ADC....156
Connecting To The Pico....158
Basic Configuration....159
The Protocol....159
SPI ADC Class....162
How Fast?....162
Summary....163

Chapter 12
Using The I2C Bus 165
I2C Hardware Basics....166
The Pico I2C....168
The I2C Functions....168
Slow Read Protocols....173
A Real Device....174
A First Program....175
The I2C Protocol In Action....177
Reading The Raw Temperature Data....177
Processing The Data....179
Reading Humidity....180
Checksum Calculation....181
Complete Listing....182
Summary....184

Chapter 13
Using The PIO 185
PIO Basic Concepts....186
State Machines And GPIO....187
PIO Blinky....188
Clock Division And Timing....190
Writing Loops....192
Data To The PIO....193
Output To GPIO....195
Side Effects....198
Input....201
Edges....203
Advanced PIO....205
Summary....207

Chapter 14
The DHT22 Sensor Implementing A Custom Protocol **209**
The DHT22....209
The Electronics....212
The Software....213
DHT22 Using the PIO – Counting....217
DHT22 Using the PIO – Sampling....223
Complete Listing....225
Summary....227

Chapter 15
The 1-Wire Bus And The DS1820 **229**
The Hardware....229
Initialization....230
Writing Bits....232
A First Command - Writing Bytes....234
Reading Bits....235
Computing The CRC....237
The DS18B20 Hardware....239
Initialization....240
Initiating Temperature Conversion....241
Reading the Scratchpad....242
Getting The Temperature....243
A Temperature Class....244
The Complete Program....245
A PIO DS18B20 Program....247
Complete Program....253
Other Commands....255
Summary....257

Chapter 16
The Serial Port **259**
Serial Protocol....259
UART Hardware....261
Setting Up the UART....262
Data Transfer....263
Serial Polling....265
Timeouts....266
Polling On Write....267
Reading and Writing Characters and Strings....269
Summary....270

Chapter 17
WiFi Using The ESP8266 **271**
The Amazing ESP8266 ... 271
Connecting the ESP8266 ESP-01 ... 272
AT Commands ... 273
Attention! ... 274
Some Utility Functions ... 277
Configuring WiFi ... 279
Connecting to WiFi ... 281
Getting A Web Page ... 282
A Web Server ... 285
Complete Listing Of Web Server ... 288
Summary ... 292

Chapter 18
Direct To The Hardware **293**
Registers ... 293
Single-Cycle IO Block ... 295
Blinky Revisited ... 296
Example I - Events ... 299
Example II PAD - Pull, Drive and Schmitt ... 301
Digging Deeper ... 304
Summary ... 305

Chapter 1
The Raspberry Pi Pico - Before We Begin

The Raspberry Pi Pico is a bold step for the Raspberry Pi world. It isn't a full System On a Chip (SoC), but a microcontroller - a small system with a minimal operating system designed to be used to control and connect to other devices. Its uses range from controlling washing machines to making games that use buttons and lights. Unlike the Raspberry Pi Zero, which is often used as a microcontroller, you can program it without having to struggle with Linux drivers and system commands. Perhaps more importantly, you don't have to fight against the multi-tasking behavior of Linux which makes getting timing right difficult. Programming the Pico means you know exactly what is happening and when, so timing becomes so much easier. It doesn't mean it's a solved problem, but it is easier. However, not having an operating system also brings its own problems and if you are used to working with the Pi Zero, say, you might well find that there are times when you would like to make a Linux call to solve a problem. Our opinion, and I hope yours too once you have read on, is that overall for the type of task the Pico is suited to, not having an operating system is a big plus point.

As to comparing the Pico to alternative microcontrollers, its big advantage is that it is powerful. It has a fast processor with enough memory to get most jobs done. It also has a great many built-in peripherals and can talk to devices such as the PWM, I2C, SPI, UART and ADC without much trouble and, with the help of the innovative PIO (Programmable I/O), just about anything else, including custom interfaces and the 1-Wire bus.

The PIO is a special feature of the Pico that is worth knowing about sooner rather than later. It is a processor in its own right and while it is restricted to only a handful of instructions it is targeted at the very specific job of working with pulsed input and output. This means that you can often write a fairly simple program to enable the PIO to code and decode any protocol that is or isn't already supported. For example, in Chapter 14 we implement a 1-Wire bus protocol and so extend the Pico to working with this popular class of device. It doesn't just stop at new protocols. For example, if you find an I2C or SPI device that doesn't quite fit into the very loose specifications of these two protocols, or goes beyond them in some way, then you can implement custom versions. Going beyond the very practical, it is worth saying that programming the PIO is great fun and very educational. It gives you an insight into how assembly language works in general.

At the moment the only negative aspect of the Pico is that it doesn't have WiFi on board. This means that, out of the box, it can't be used as an IoT (Internet of Things) device as it can't connect to the Internet. However, this is easily fixed. In Chapter 17 we discover that it is simple and cheap to add an ESP8266 WiFi device via a single serial interface.

A very important point about the Pico is that it is widely available. This might seem like a strange thing to say, but the Raspberry Pi Zero is still restricted to one per customer and, as such, not really a good candidate for moving from a prototype to a commercial product. The Pico can be bought in any quantity and so can the chip it is based on. As such, it is suitable for a project that might graduate to a fully finished end product.

This is a sketchy overview of what the Pico is all about. In the rest of the chapter we look at the details of the Pico and the things you need to get started.

The Pico And Its Family

The Pico is based on the RP2040 processor chip which was designed by Raspberry Pi specially for the Pico. It is based on the ARM Cortex M0 and runs ARM machine code. The processor is available as a chip that can be purchased to build your own implementation of a Pico-like device. Many other well-known hardware companies, notably Arduino, have done just this and, if you want a Pico, you aren't restricted to just the Raspberry Pi Pico, there are other choices. Some are very small and some have extras on board:

These alternatives to the Pico are easy to use because they operate just like the Pico. You can program them using MicroPython and the same environment. The only differences might be that the smaller form factors expose fewer hardware lines so you might not be able to use everything you can use on the full Pico. In the remainder of this book the programs target the Raspberry Pi Pico, but they should just work or be easily modifiable for any device that used the RP2040.

The key points about the Pico hardware are:

- Dual-core Arm Cortex M0+ processor, flexible clock running up to 133 MHz
- 264KB of SRAM, and 2MB of on-board Flash memory
- USB 1.1 with device and host support
- Low-power sleep and dormant modes
- Drag-and-drop programming using mass storage over USB
- 26 × multi-function GPIO pins
- 2 × SPI, 2 × I2C, 2 × UART, 3 × 12-bit ADC, 16 × controllable PWM channels
- Accurate clock and timer on-chip
- Temperature sensor
- Accelerated floating-point libraries on-chip
- 8 × Programmable I/O (PIO) state machines for custom peripheral support

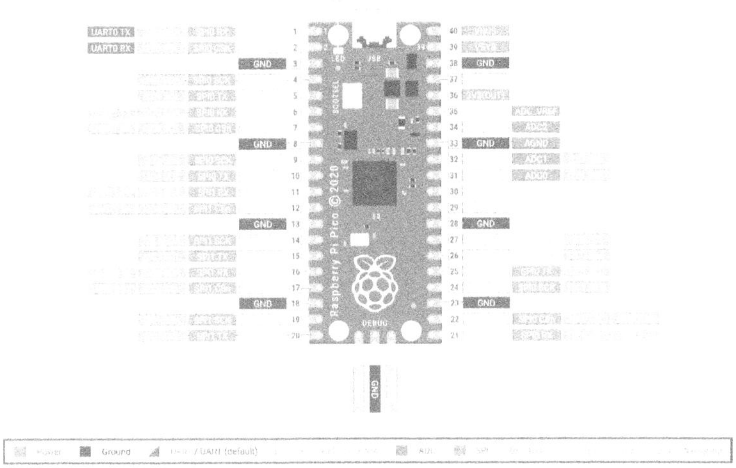

You could also consider the on-board LED, which is indicated on the diagram, as a key feature – it may be small, but it is very useful in getting started and testing.

What To Expect

There are no complete projects in this book – although some examples come very close and it is clear that some of them could be used together to create finished projects. The reason for this is that the focus is on learning how things work so that you can move on and do things that are non-standard.

What matters is that you can reason about what the processor is doing and how it interacts with the real world in real time. This is the big difference between desktop and embedded programming. In the desktop world you don't really care much about when something happens, but when you are programming a physical system you care very much.

This is a book about understanding general principles and making things work at the lowest possible level. This knowledge isn't always necessary when you are working on a relatively slow system in, say, MicroPython, but it is always helpful for understanding what is going on when things go wrong. When you are working directly with the hardware, speed matters and so does knowing what is happening.

All of the examples are as basic as possible and the code is designed to be as easy to understand as possible. In most cases this means avoiding the use of constants that appear to come from nowhere and functions that make it difficult to see the basic steps. Also error handling is reduced to a bare minimum – simple programs look complicated if you add error handling code. Of course, there is no reason not to refactor these examples into something that looks more like production code and the effort in doing this is much less than getting the basic programs working in the first place.

Rather than go through multiple possible configurations for a development environment, this book uses Thonny and VS Code from the start. If you want to work in a different way then the documentation has instructions.

What Do You Need?

Well – a Pico at least! In fact you would be well advised to buy at least two Picos – accidents happen. The Pico is small and lacks an operating system, so running all of the software you need to create programs on it needs another machine – the development machine. The good news is that you can use almost any desktop machine – PC, Mac or Linux system. If you want to use a Raspberry Pi then a Pi 4 or Pi 400 is recommended as running development software needs a reasonably powerful machine.

As to additional hardware over and above the Picos, you will need a solderless prototype board and some hookup wires – known as Dupont wires. You will also need some LEDs, a selection of resistors, some 2N2222 or other general purpose transistors and any of the sensors used in later chapters. See the Resources page for this book on the I/O Press website for links. It is probably better to buy what you need as you choose to implement one of the projects, but an alternative is to buy one of the many "getting started" kits. You will probably still need to buy some extra components, however.

A solderless prototype board and some Dupont wires

While you don't need to know how to solder, you will need to be able to hook up a circuit on a prototyping board. A multimeter (less than $10) is useful, but if you are serious about electronic projects, investing in a logic analyzer (less than $100) will repay itself in no time at all. You can get small analyzers that plug in via a USB port and use an application to show you what is happening. It is only with a multichannel logic analyzer that you have any hope of understanding what is happening. Without one and the slight skill involved in using it, you are essentially flying blind and left to just guess what might be wrong.

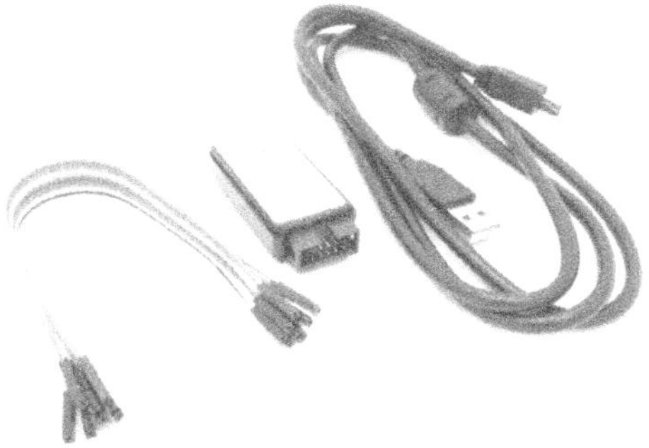

A Low Cost Logic Analyzer

Finally, if you are even more serious, then a pocket oscilloscope is also worth investing in to check out the analog nature of the supposedly digital signals that microcontrollers put out. However, if you have to choose between these two instruments, the logic analyzer should be your first acquisition.

It is worth noting that the Pico can generate signals that are too fast to be reliably detected by low-cost oscilloscopes and logic analyzers, which work at between 1MHz and 25MHz. This can mean that working with pulses much faster than $1\mu s$ can be difficult as you cannot rely on your instruments. There are reasonably priced 200MHz and 500MHz logic analyzers and one of these is certainly worthwhile if you are serious about hardware. It is worth knowing that both instruments can mislead you if you try to work with signals outside of the range that they can work with.

It is also assumed that you are able to program in Python. While there are some differences between it and MicroPython, the programs are easy enough to follow and any out-of-the-ordinary coding is explained.

Community

If you get stuck the best place to ask for help is in the Pico forums or, second best, Stack Overflow. The quality of answers varies from misleading to excellent. Always make sure you evaluate what you are being advised in the light of what you know. Be kind and supportive of anyone offering an answer that indicates that they misunderstand your question.

You also need to keep in mind that the advice is also usually offered from a biased point of view. Experts in other language will often give you a solution that abandons MicroPython. Electronics beginners will offer you solutions that are based on “off-the-shelf” modules, when a simple alternative solution is available, based on a few cheap components. Even when the advice you get is 100% correct, it still isn’t necessarily the right advice for you. As a rule never follow any advice that you don’t understand.

Summary

- The Pico is a remarkably powerful device given its low cost and is ideal for building prototypes, one-offs and production devices.
- The RP2040 chip that is at the heart of the Pico is also used in a number of similar devices, all of which can be treated as variations on the basic Pico.
- You need to decide on a development machine to use in conjunction with your Pico. A PC, Mac or a Linux system are all suitable, including a Pi 4 or a Pi 400.
- Thonny or VS Code provide an easy-to-use and efficient development environment, irrespective of the type of development machine you choose.
- To work with electronics, you need a solderless prototyping board, some hookup wires and some components. You also need a multimeter and preferably a logic analyzer. After these basic instruments you can add what you can afford.
- There is an active Pico community forum and if you get stuck it's the place to ask for advice. However, always evaluate any advice proffered and, in general, don't accept it unless you understand it.

Chapter 2
Getting Started

The easiest language to use to program the Pico is MicroPython. This is a reasonably full implementation of Python 3 plus special modules to work with the Pico's hardware. If you know the Python language you will have no problem working in MicroPython. However, getting used to the ideas involved in working with hardware is another matter – you have to think a little differently. To put it simply, time matters. What this means will become clear in the rest of the book, but exactly when and in what order things happen are fundamental concerns to this sort of hardware programming, and this usually means needing the most efficient programming language possible.

Sometimes, however, you don't need speed, even in an IoT application. For example, if you just want to flash a few LEDs or read a temperature sensor in a human timescale, then you can write in almost any language and MicroPython is ideal. You can also often avoid having to react at the highest possible speed by using the range of peripheral devices that the Pico has. In other words, you can offload time critical operations to specialized hardware.

Speed, or rather lack of speed, can be a problem with coding in Python, but it is worth explaining that while Python may not be fast, it is sophisticated. There are ways of writing code in MicroPython that would require a lot of work to implement in lower-level languages. You may not have raw speed, but you do have mature sophistication.

So MicroPython is worth learning and the Pico provides low-cost hardware to experiment with. How do we get started?

Installing MicroPython

The key to understanding how everything works is to realize that what we are about to do is convert a Pico into a MicroPython machine. That is, we are going to download the MicroPython system onto a Pico and from this moment on the Pico behaves quite differently because every time you switch it on it is running the MicroPython system.

We get the MicroPython program onto the Pico using the basic way of getting any program onto the Pico. The difference is that we are only going to do this once. If nothing goes wrong after we have installed MicroPython we can use it to download and run any MicroPython programs we write in future. You only have to repeat the installation if something damages the MicroPython system or you load some other program onto the Pico.

The basic Pico way of installing programs is to simply drag-and-drop a `.uf2` format file to a folder on the Pico. First you need to download the MicroPython uf2 file from the Raspberry Pi website. At the time of writing the file is called:

`rp2-pico-20210205-unstable-v1.14-8-g1f800cac3.uf2`

but this is likely to change.

To run the program the simplest thing to do is connect a Pico to the development machine, a PC, Mac or a Raspberry Pi, via USB. The USB connection will also power the Pico, and to get the Pico to present a USB drive to the development machine you have to hold down the `BootSel` button which is to the left of the USB connector as the power is applied. If you have done this correctly you will see a new drive added to the development machine and you will have to supply a password to mount it. Under Linux, for example, you will see something like:

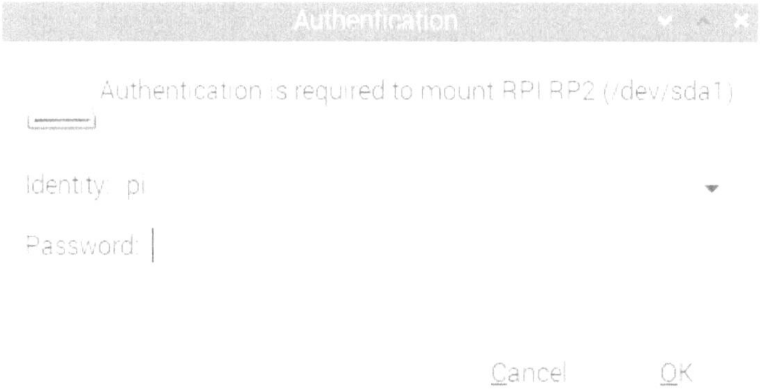

If you don't see this dialog box then you haven't held down the `BootSel` button while the power is being applied. Now you can drag-and-drop the MicroPython `uf2` file and you should see the on-board LED start to flash slowly.

When the transfer is complete the Pico USB drive will automatically disconnect and you can remove the power from the Pico or start using it at once. The point is that now whenever you power on that particular Pico it will run as a MicroPython machine. If you don't want it to do this you need to download another program using the same method.

There are various ways of checking that the program has been successfully loaded into the Pico, but the process is so simple that it rarely goes wrong and it is worth moving on to a more sensible way of working with the MicroPython Pico.

Hello World Using Thonny

As we have downloaded the MicroPython code into the Pico we are ready to try out a first program which by tradition is "Blinky", i.e. a program to flash an LED. We could do this via the command prompt, but it is much more efficient to use an IDE. The IDE of choice for the Pico is Thonny which you will find already installed on most Raspberry Pis. It can also be installed under Windows, Mac or Linux, however. If you already have it installed you need to check that it is up to date. Visit the Thonny website:

```
https://thonny.org/
```

and download the latest version.

If you are working with Linux or a Pi you can use:

```
sudo apt install thonny
```

Once installed, start it running and make sure that the Pico with MicroPython is connected via a USB cable, as Thonny and other IDEs communicate with the Pico via the serial port provided by the USB cable. You can connect to the Pico via an on-board UART, but this is complicated and unnecessary when you are first getting started.
As long as the Pico is connected, and no other program is making use of its serial connection, Thonny should be able to find it automatically. Select the Tools, Options menu item and then the Interpreter tab in the dialog box:

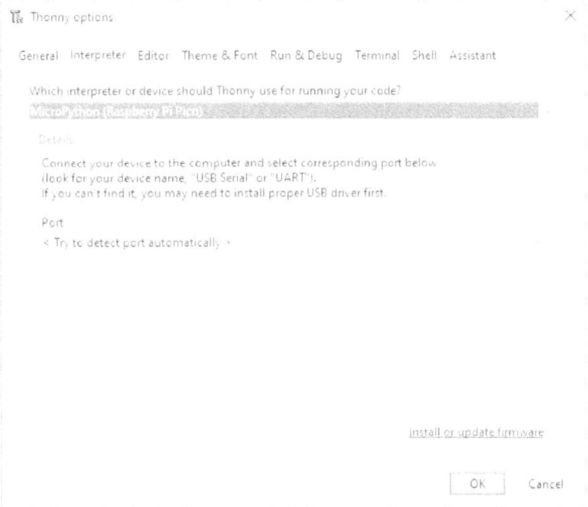

Select the MicroPython (Raspberry Pi Pico) option and choose the <try to detect port automatically> option. If it fails try setting the port that you know the Pico is connected to.

As long as this works you will see an editing window and a Shell that you can type immediate MicroPython instructions into and see the commands that Thonny issues to the Pico.

Enter the following program:

```
print("Hello Pico World")
```

This will print the message to the Pico console at the bottom of the Thonny window. Once the program is entered you can click the Run icon and select where you want the program to run – locally or on the Pico:

If you select the Pico, you will be asked to provide a name for the program and it will be downloaded to the Pico and run. You should see the confirmation message appear.

MicroPython creates a small filing system on the Pico and you can save and open files stored there using Thonny.

Thonny is a good way to get started as it is simple and quick, but it lacks many of the facilities you may be used to in other IDEs. In particular it lacks any debug option and it has no intellisense or syntax checking for MicroPython.

Hello World VS Code

There are some advantages to using VS Code to work with MicroPython on the Pico, but it is slightly more difficult to set up as you need to install a number of extensions. The main advantage is that you can make use of intelligent prompting when entering code and the Pylance linter to spot simple syntax problems. It also has a more comprehensive way of working with the Pico's filing system.

First you need to install VS Code. This is not difficult and instructions on how to do this on Windows, Mac and Linux are available on the VS Code website and there isn't any need to repeat them here. Make sure you have VS Code installed and working before moving on to the next step.

You also need to install Python 3, and again there are good instructions on the website. Make sure that Python 3 is installed and working before you move on.

The final steps are installing the required VS Code extensions. You need to install the Microsoft Python Extension:

Next install Pylance:

Finally install the Pico-Go extension:

Now you should be ready to create and run MicroPython Programs. Create a folder called `PicoPython` or something similar and open it in VS Code. Next create a file called `hello.py` containing:

```
print("Hello Pico World")
```

If you can't see the extension tools in the status bar then restart VS Code and open the folder. If you can see the tools in the status bar then, as long you can also see Pico Connected, you are ready to run the program:

To do this, simply click the Run command in the status bar. After a moment you should see the message appear in the Pico Console.

You can see other commands in the status bar and the All Commands option at the end gives you access to even more:

The Configure project command should be run at least once on any project as it adds auto-complete and linting (error checking) It also adds `pico-go.json` file to `.vscode` which you can use to change defaults. You can edit this file by using Global Settings.

If you need access to the filing system on the Pico use the Start FTP server command. Once the server has started you can connect to it using any FTP client. Under Windows, for example, you can open a command prompt and enter the ftp command followed by:

```
ftp> open 127.0.0.1 2121
Connected to 127.0.0.1.
220-Pico FTP - welcome!
220 Ready
530 Command requires authentication: OPTS
User (127.0.0.1:(none)):
```

The user name and password is pico. From here you can use the standard ftp commands to copy files from the local machine to the Pico.

Summary

- MicroPython is a powerful and sophisticated language and, with the Pico's speed, you can achieve a great deal.
- Before you can start using MicroPython, you have to download it and install it on any Pico you want to use.
- Installing it onto a Pico is just a matter of connecting the Pico to the development machine, opening the Pico as a folder, and drag-and-dropping the MicroPython uf2 file onto it.
- Once MicroPython is installed on a Pico it is a MicroPython machine every time you switch it on.
- To program a MicroPython machine you can use the command prompt, but it is so much easier to use an IDE.
- The standard IDE for Pico MicroPython is Thonny and this works with the Pico without any additional configuration.
- Thonny is very good, but VS Code has the advantage of intellisense prompting and error checking and is very easy to install.
- The VS Code Pico-Go extension has many additional commands that let you work with the file system on the Pico.

Chapter 3
Getting Started With The GPIO

In this chapter we take a look at the basic operations involved in using the Pico's General Purpose Input/Output (GPIO) lines with an emphasis on output. We'll consider questions such as how fast can you change a GPIO line, how do you generate pulses of a given duration and how can you change multiple lines in sync with each other?

Pico Pins

The first thing to make ourselves familiar with is the layout and range of GPIO pins available on a standard Pico – some compatible devices have fewer or differently arranged pins. The RP2040 chip has 30 GPIO lines, GP0 to GP29, of which the Pico makes 26 available for general use on the edge of the board. The other four are used internally. GP23 is used as a power control line, GP24 is used to sense the USB power input, GP25 powers the on-board LED and GP29 is used to measure the system voltage. There are also four additional GPIO lines, but these are dedicated for use as a quad SPI bus to interface to flash memory.

As is the case with most microprocessors, each GPIO line, with the exception of GP22, has multiple uses as you can see in the diagram on the next page.

You can select what mode a pin is used in and in this chapter we concentrate on using pins in the simplest GPIO mode. Even so, which pins you select for general-purpose use should take into account what other uses you might put pins to. For example, using GP26 as a general GPIO line when you need an additional ADC (Analog to Digital Converter), which is one of its alternative functions. The only pin that has no alternative function is GP22 and this should be your first choice for any general-purpose duties.

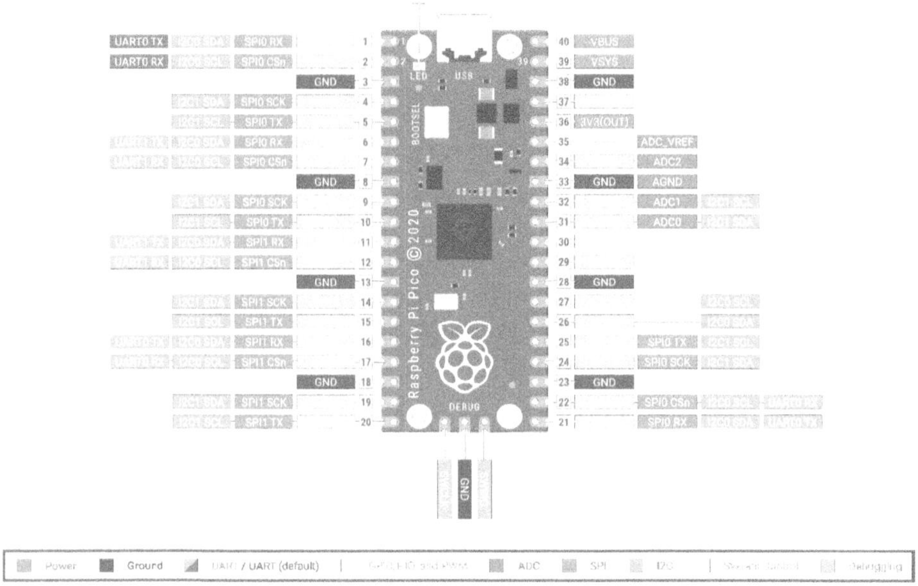

Notice that, unlike when programming directly to the hardware, MicroPython doesn't use the idea of setting the GPIO pin into a particular mode. Instead it provides classes that make use of GPIO pins in particular ways. For example, if you were programming directly to the hardware you would first set the pins you wanted to use to the mode you wanted to use, e.g. PWM (Pulse Width Modulation) and then you would start working with PWM operations. In MicroPython you would simply create a PWM object using the pin in question and expect it to take care of setting the pin to the correct mode. Notice, however, that you are still restricted to using pins that support the mode you are using.

In most cases, to make use of the GPIO lines in a prototyping situation you need to solder pins onto the board so that it can be plugged into a prototype board. There is some argument about which side the pins should be soldered to. One side leaves the PCB legend visible, but the on-board LED is obscured, and the other leaves the LED visible, but obscures the legend.

Basic GPIO Functions

The MicroPython `Pin` class allows you to create an object which controls the way a single GPIO line works. The simplest form of the constructor is:

```
Pin(id, mode=mode)
```

where *id* is the number of the GPIO line you want to use and *mode* is one of:

```
Pin.IN
```

or

```
Pin.OUT
```

there are other possibilities and these are discussed later.

There is also an `init` method which can be used to change the configuration of the pin. For example,

```
pin.init(mode=Pin.IN)
```

might be used to change a pin from output to input.

A number of methods are provided to work with the state of a Pin object:

Method	Description
`value(x)`	Sets the line to x, usually 0 or 1, but x can be anything that evaluates to true or false
`on()`	Sets the line to high
`off()`	Sets the line to low
`high()`	Sets the line high
`low()`	Sets the line low
`toggle()`	If the line is high set it low and vice versa

There are a number of other methods and properties but these are the most basic.

Blinky

The first IoT program you write is by tradition Blinky, i.e. flash an LED. As the Pico has an on-board LED connected to GP25 all ready to flash, the simplest Blinky program is:

```
from machine import Pin
import time

pin = Pin(25, Pin.OUT)

while True:
    pin.value(1)
    time.sleep(1)
    pin.value(0)
    time.sleep(1)
```

If you run this program you should see the on-board LED flash at 1-second intervals. The program simply sets GP25 to output and then uses the value method to change its state from high to low with a suitable delay, provided by `sleep`.

This was very easy, but a real program to flash an LED should use a general I/O line and an external LED. With this in mind, let's flash an LED connected to GP22.

Enter the program:

```
from machine import Pin
import time

pin = Pin(22, Pin.OUT)
while True:
    pin.toggle()
    time.sleep(1)
```

The difference is that now we are using GP22 and the toggle method to change the state of the line. The program also doesn't use any constants in order to make what is happening clearer. It first initializes GP22 to be an output and sets it repeatedly high and low.

If you want to connect an LED to see the "blinking" for real then this is easy enough, but you do need a current-limiting resistor to avoid the LED drawing more current than the Pico GPIO line can supply and possibly damaging the chip. A 200Ω resistor is a good choice, see Chapter 5. A better way to drive an LED is discussed more fully in Chapter 5.

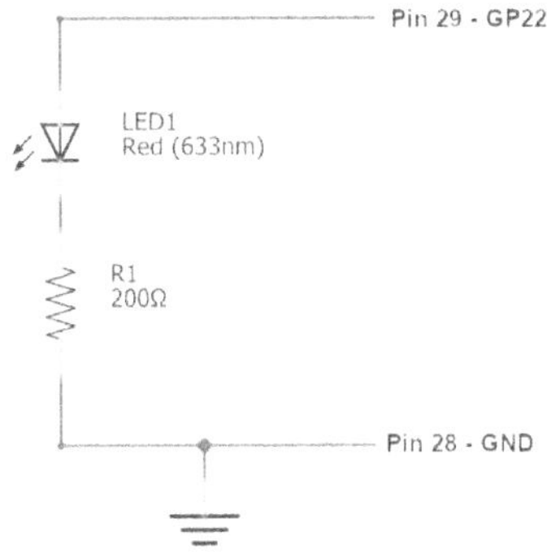

How you build the circuit is up to you. You can use a prototyping board or just a pair of jumper wires. The short pin and/or the flat on the side of the case marks the negative connection on the LED – the one that goes to ground.

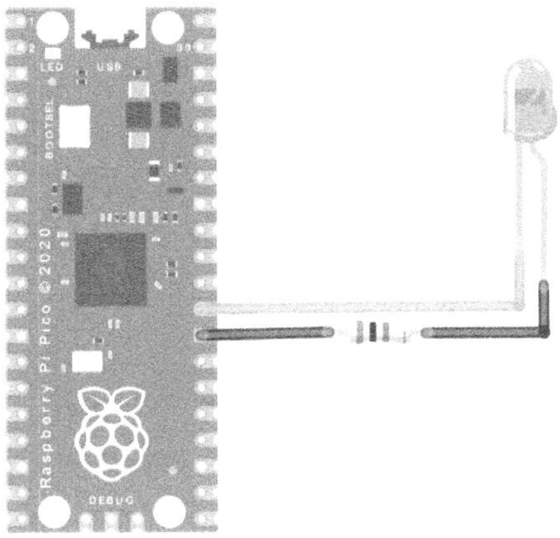

If you can't be bothered to go though the ritual of testing "Blinky" with a real LED, then just connect a logic analyzer to Pin 7 and you will see one-second pulses.

Summary

- The Pico has 34 GPIO lines in total. Four are used to interface to flash memory and of the remainder, GP0 to GP29, the Pico makes 26 available for general use on the edge of the board.

- Some of the GPIO lines have alternative functions and these are best avoided if all you need is a simple GPIO input/output line.

- The only GPIO line that doesn't have another important function is GP22 and this is your best first choice for a general use GPIO line.

- MicroPython provides the Pin class to control a single GPIO line and its basic methods let you set the line high or low.

- A Blinky program to flash an LED is trivial given that the Pico has an on-board LED connected to GP25 all ready to flash, but if you want to test Blinky with an external LED on another GPIO line you need to include a current-limiting resistor in the circuit to avoid the LED drawing more current than the Pico GPIO line can supply.

Chapter 4

Simple Output

A GPIO line is either configured to be an input or an output. The electronics of working with inputs and outputs are discussed in the next chapter, but first we focus on the software side of the task of using GPIO lines in output mode. While it isn't possible to ignore electronics entirely, keep in mind that more in details are provided in Chapter 5.

It is worth noting at this stage that output is easy. Your program chooses the time to change a line's state and you can use the system timer to work out exactly when things should happen. The real problems only start to become apparent when you are trying to change the state of lines very fast or when they need to be changed synchronously. This raises the question of how fast the Pico can change a GPIO line and this is something we consider at this early stage because it puts constraints on what we can easily do.

Basic GPIO Functions

We have already met the basic methods of the Pin object that let you work with a single GPIO line:

Method	Description
`init(mode)`	Set mode to input or output
`value(x)`	Sets the line to x, usually 0 or 1, but x can be anything that evaluates to true or false
`on()`	Sets the line to high
`off()`	Sets the line to low
`high()`	Sets the line high
`low()`	Sets the line low
`toggle()`	If the line is high set it low and vice versa

Using these methods is very straightforward, but notice that there is no way to set multiple lines in one operation. This is can be a problem, something we'll come to later.

How Fast?

A fundamental question that you have to answer for any processor intended for use in embedded or IoT projects is, how fast can the GPIO lines work?

Sometimes the answer isn't of too much concern because what you want to do only works relatively slowly. Any application that is happy with response times in the tens of millisecond region will generally work with almost any processor. However, if you want to implement custom protocols or anything that needs microsecond, or even nanosecond, responses, the question is much more important.

It is fairly easy to find out how fast a single GPIO line can be used if you have a logic analyzer or oscilloscope. All you have to do is run the program:

```
from machine import  Pin
pin = Pin(22, Pin.OUT)
while True:
    pin.value(1)
    pin.value(0)
```

If you run this program you will discover that the pulses are about 8μs and not perfectly even:

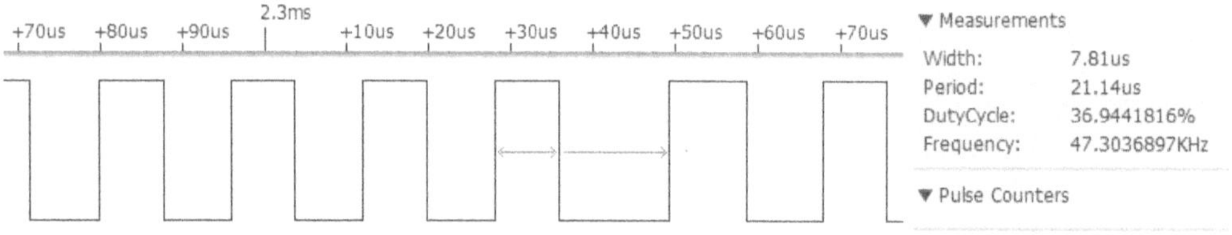

The unevenness is due to the internal workings of MicroPython. If you change the way that the code is specified, you are likely to see changes in timing. For example, if you try the equivalent code:

```
def flash():
    pin = Pin(22, Pin.OUT)
    while True:
        pin.value(1)
        pin.value(0)
flash()
```

then you will discover that the pulse width has dropped to around 5μs.

There is a facility to compile functions to native code and this gives the largest speed increase without going to exceptional lengths. If you want a function to be complied all you have to do is add the `@micropython.native` decorator:

```
from machine import Pin

@micropython.native
def flash():
    pin = Pin(22, Pin.OUT)
    while True:
        pin.value(1)
        pin.value(0)
flash()
```

If you run this program you will find that, not only does the pulse time drop to 3.8μs, but the pulses are very regular.

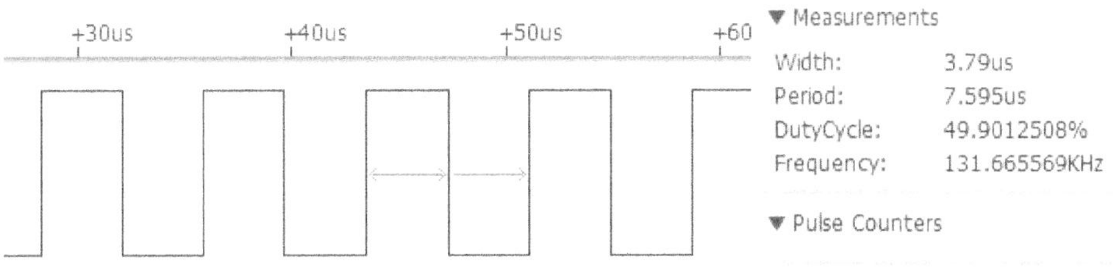

The reason the pulses are regular is that now the MicroPython system doesn't get involved in the `while` loop and it runs at the same speed all of the time. One problem with this approach is that, as the native function now doesn't give MicroPython a chance to run, you lose control of the Pico via the USB connection. You now have to reset to gain control of the Pico. You should restrict the use of native code as much as possible and only use it when speed is unattainable by other methods.

As frequently mentioned in this book, when speed is an important factor the programming language of choice is C and this provides a good example of why. If you compare the approximately 4μs pulses that can be produced using MicroPython on the Pico to the 50ns pulses that can be produced using C, you can appreciate that MicroPython is about 100 time slower than C.

Including Pauses

To generate pulses of a known duration we need to pause the program between state changes. In the Blinky programs we used `sleep` to slow things down, but without properly introducing it.

Using `sleep(seconds)` gives a pause or "wait" for the specified number of seconds. As *seconds* is a floating point number you can specify fractions of a second. So for half-second pulses you could use:

```
from machine import Pin
import time

pin = Pin(22, Pin.OUT)
while True:
    pin.value(1)
    time.sleep(0.5)
    pin.value(0)
    time.sleep(0.5)
```

As well as `sleep` there are also `sleep_ms` and `sleep_us` which pause the program for the specified number of milliseconds and microseconds respectively. You can use `utime` or `time` to import the functions.

Of course, when creating pulses of a given time the waits add to the basic pulse time. So for example:

```
from machine import Pin
import time

pin = Pin(22, Pin.OUT)
while True:
    pin.value(1)
    time.sleep_us(10)
    pin.value(0)
    time.sleep_us(10)
```

Creates pulses that are $26\mu s$ wide not $10\mu s$. In general, you have to add about $10\mu s$ to the wait time to get the pulse length.

The traditional way of introducing a busy wait (also known as a spin wait) is to simply use a time-wasting `for` loop. A for loop busy wait can produce short wait times:

```
pin = Pin(22, Pin.OUT)
n = 10
while True:
    for i in range(n):
        pass
    pin.value(1)
    for i in range(n):
        pass
    pin.value(0)
```

which generates pulses according to the setting of `n`:

n	Time in μs
1	24-30
2	30-38
3	38-44
4	46-52
5	54-60
6	62-70
7	70-76
8	82-84
9	90-94
10	98-100

Fixed Time Delay

A common problem is making sure that something happens after a fixed time delay when you have a variable amount of work to do during that time interval. Consider the program snippet:

```
pin.value(1)
for i in range(n):
   pass
time.sleep_ms(1)
pin.value(1)
```

where the `for` loop is intended to stand in for doing some other work. The intention is that the GPIO line should be set high for 1ms, but clearly how long the line is set high depends on how long the loop takes, which is given by `n` plus 1ms of sleep time.

What is needed is a pause that takes into account the time that the loop uses up and simply delays the program for the remaining amount of time to make it up to 1ms. This is where the functions `ticks_ms()` and `ticks_us()` come in useful. They give the time since the machine was switched on in milliseconds or microseconds respectively. These both wrap around at some unspecified point and to take account of the wrap you need to use:

```
ticks_add(ticks,number)
ticks_diff(ticks1,ticks2)
```

to do arithmetic that takes account of the wrap.

We can now write the program snippet given earlier as:

```
while True:
    t=time.ticks_add(time.ticks_us(),1000)
    pin.value(1)
    for i in range(n):
        pass

    while time.ticks_us()<t:
        pass
    pin.value(0)
```

Now we obtain the ticks before setting the line high and add 1000 to it. No matter how long the for loop takes, the while loop will provide a delay of 1ms, as long as the loop takes less than this time.

This is a very general technique and one that can often make difficult timing problems very simple.

Phased Pulses

As a simple example of using the output functions, let's try to write a short program that pulses two lines, high and then low, out of phase.

The simplest program to do this job is:

```
from machine import Pin

pin1 = Pin(21, Pin.OUT)
pin2 = Pin(22, Pin.OUT)
while True:
    pin1.value(1)
    pin2.value(0)
    pin1.value(0)
    pin2.value(1)
```

Notice that there is no delay in the loop so the pulses are produced at the fastest possible speed.

Using a logic analyzer reveals that the result isn't what you might expect:

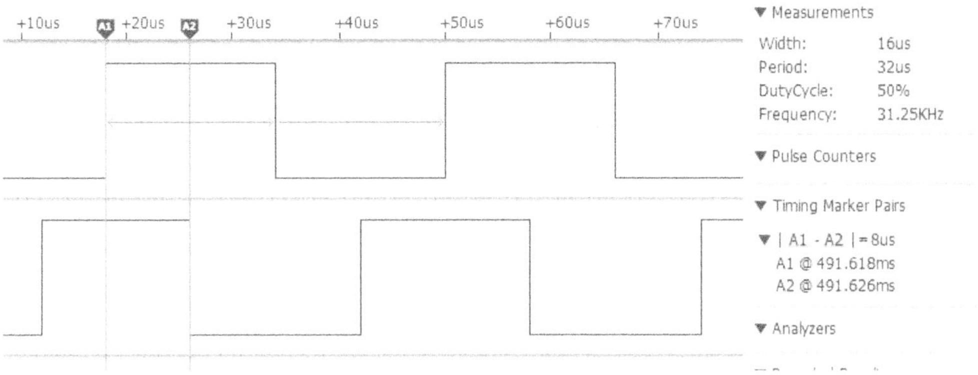

The top train switches on and the bottom train takes about half a pulse before it switches off, although the intent is for both actions to occur at the same time. The point is that it does take quite a long time to access and change the state of an output line.

Of course, if we include a delay to increase the pulse width then the delay caused by accessing the GPIO lines in two separate actions isn't so obvious, but it is still there. There are applications where the switching speed is so low that the delay between switching doesn't matter – flashing LEDs for instance. With a delay of around 8μs you could flash a line of around 2000 LEDs before the lag between the first and the last became apparent. On the other hand, if you use out-of-phase pulses to control a motor, then the overlap when both GPIO lines were on would burn out the drivers quite quickly. Of course, any sensible, cautious, engineer wouldn't feed a motor control bridge from two independently generated pulse trains unless they were guaranteed not to switch both sides of the bridge on at the same time.

Setting Multiple GPIO Lines

There is no way using MicroPython methods to change multiple GPIO lines at the same time, even though the hardware makes it possible. To do the job you need to write some code that accesses the hardware directly – see Chapter 18 for more details.

In this chapter we simply present and make use of the two functions explained there:

```
def gpio_get():
    return machine.mem32[0xd0000000+0x010]

def gpio_set(value,mask):
 machine.mem32[0xd0000000+0x01C]=
                 (machine.mem32[0xd0000000+0x010])^value & mask
```

Both functions work by directly accessing the GPIO registers.

The first, `gpio_get`, returns a 32-bit word that has a single bit for the current state of each of the GPIO lines. That is, it reads all of the GPIO lines in a single operation.

The second function, `gpio_set`, uses a mask to determine which lines will be set and a value that gives the states to set them to. As in the case of `gpio_get`, each line is represented by a single bit in the value and the mask. Any bits not set in the mask leave the corresponding GPIO line unchanged.

It is easy to create a mask for any GPIO lines. For example if you want to modify only lines GP`n` and GP`m` then the mask is:

```
mask=1<<n | 1<<m
```

and so on if you have more lines to modify.

The value can be constructed in the same way. If you want to set the lines to `a` and `b` then the value is:

```
mask=a<<n | b<<m
```

Notice that if the corresponding bit isn't set in `mask` then the bit in `value` has no effect.

Making use of this we can write the previous program without the lags:

```
from machine import Pin
import machine

def gpio_get():
    return machine.mem32[0xd0000000+0x010]

def gpio_set(value,mask):
    machine.mem32[0xd0000000+0x01C]=
                 (machine.mem32[0xd0000000+0x010])^value & mask


pin=Pin(22,Pin.OUT)
pin=Pin(21,Pin.OUT)
value1=1<<22 | 0<<21
value2=0<<22 | 1<<21
mask=1<<22 | 1<<21
while True:
    gpio_set(value1,mask)
    gpio_set(value2,mask)
```

As we are changing the same pins each time, we only need a single mask. The value, however, changes each time. If you run this program you will see an almost perfect pair of out-of-phase 44μs pulses:

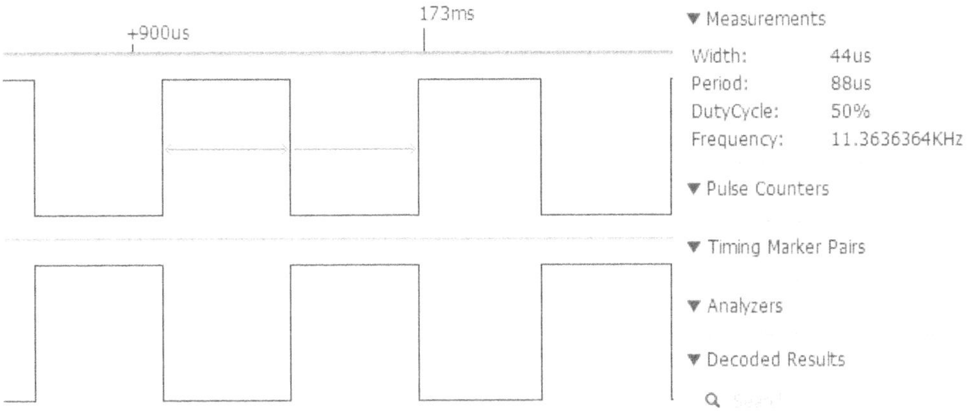

Summary

- Output is easy because the program decides when to change the state of a line. Input is hard because you never know when an input line will change state.
- GPIO lines can be set to act as inputs or outputs when you create a Pin object.
- If a line is set to output it can be set high or low using a number of Pin Methods function.
- You can generate pulses as short as $8\mu s$, but not reliably.
- If you compile a function that changes the GPIO lines you can generate reliable $4\mu s$ pulses
- A delay can be introduced into a program using the `sleep, sleep_ms` or `sleep_us` functions.
- An alternative is to use a busy wait loop which is simply a loop that keeps the CPU busy for an amount of time. It is easy to obtain an equation that gives the delay per loop repetition.
- By using the tick methods you can set an action to have an exact repeat time, even if what it does varies in time.
- Producing pulses which are in accurately in phase is not possible using Pin methods.
- If you access the hardware directly you can change multiple lines in one operation.

Chapter 5
Some Electronics

Now that we have looked at some simple I/O, it is worth spending a little time on the electronics of output and input. We cover the electronics of input before looking at how the software handles input because we need to understand some of the problems that the software has to deal with.

First some basic electronics – how transistors can be used as switches. The approach is very simple, but it is enough for the simple circuits that digital electronics makes use of. It isn't enough to design a high-quality audio amplifier or similar analog device, but it might be all you need.

The basis of all electronics is Ohm's law, V = IR, and this prerequisite implies an understanding of voltage, current and resistance.

Electrical Drive Characteristics

If you are not familiar with electronics, the important things to know are what voltages are being worked with and how much current can flow. The most important thing to know about the Pico is that it works with two voltage levels – 0V and 3.3V. The RP2040 chip can work at a range of voltages from 1.8V to 3.3V, but the Pico has a power supply that converts whatever you supply it with to 3.3V – as a result the Pico is a 3.3V logic device.

If you have worked with other logic devices you might be more familiar with 0V and 5V as being the low and high levels. The Pico uses a lower output voltage to reduce its power consumption, which is good, but you need to keep in mind that you may have to use some electronics to change the 3.3V to other values. The same is true of inputs, which must not exceed 3.3V or you risk damaging the Pico.

An important question is how much current the GPIO lines can handle without damaging the chip. This isn't an easy question and at the time of writing the documentation isn't clear on the matter. According to the documentation, each GPIO line can be set to "drive" 12mA. However, this doesn't quite mean what you might think. This is not an upper limit on the supplied current, but a configuration that is needed to ensure that the output voltages of the GPIO line are within specification while it is working at 12mA – see Chapter 18 for more information.

What this means is that the designers intended the GPIO line to be used at 12mA, but this is not an upper limit on supply current. The only upper limit quoted is that the total current in the GPIO lines should be less than 50mA. Given that there are 30 GPIO lines this gives an average of 1.6mA per GPIO line. In practice, you are most likely to be safe at around the 12mA maximum for a small number of lines.

In practice, if you are planning to use more than 1.6mA from multiple GPIO lines, consider using a transistor. If your circuits draw more than 50mA from the 3.3V supply rail, consider a separate power supply.

Notice that the 12mA limit means that you cannot safely drive a standard 20mA red LED without restricting the current to below 12mA. A better solution is to use a low-power 2mA LED or use a transistor driver.

Driving An LED

One of the first things you need to know how to do is compute the value of a current-limiting resistor. For example, if you just connect an LED between a GPIO line and ground then no current will flow when the line is low and the LED is off, but when the line is high, at 3.3V, it is highly likely that the current will exceed the safe limit. In most cases nothing terrible will happen as the Pico's GPIO lines are rated very conservatively, but if you keep doing it eventually something will fail. The correct thing to do is to use a current-limiting resistor. Although this is an essential part of using an LED, it is also something you need to keep in mind when connecting any output device. You need to discover the voltage that the device needs and the current it uses and calculate a current-limiting resistor to make sure that is indeed the current it draws from the GPIO line.

An LED is a non-linear electronic component – the voltage across is stays more or less the same irrespective of the current passing through the device. Compare this to a more normal linear, or "ohmic", device where the current and voltage vary together according to Ohm's law, V =IR, which means that if the current doubles, so does the voltage and vice versa.

This is not how an LED behaves. It has a fairly constant voltage drop, irrespective of the current. (If you are curious, the relationship between current and voltage for an LED is exponential, meaning that big changes in the current hardly change the voltage across the LED.) When you use an LED you need to look up its forward voltage drop, about 1.7V to 2V for a red LED and about 3V for a blue LED, and the maximum current, usually 20mA for small LEDs. You don't have to use the current specified, this is the maximum current and maximum brightness.

To work out the current-limiting resistor you simply calculate the voltage across the resistor and then use Ohm's law to give you the resistor you need for the current required. The LED determines the voltage and the resistor sets the current.

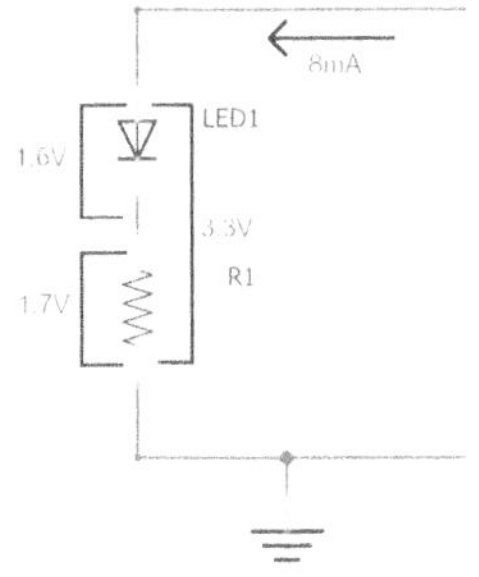

A GPIO line supplies 3.3V and if you assume 1.6V, its forward voltage, across the LED that leaves 1.7V across the current-limiting resistor since voltage distributes itself across components connected in series. If we restrict the current to 8mA, which is very conservative, then the resistor we need is given by:

R = V/I = 1.7/8 = 0.212

The result is in kiloohms, kΩ, because the current is in milliamps, mA. So we need at least a 212Ω resistor. In practice, you can use a range of values as long as the resistor is around 200 ohms – the bigger the resistor the smaller the current, but the dimmer the LED. If you were using multiple GPIO lines then keeping the current down to 1 or 2mA would be better, but that would need a transistor.

You need to do this sort of calculation when driving other types of output device. The steps are always the same. The 3.3V distributes itself across the output device and the resistor in some proportion and we know the maximum current – from these values we can compute the resistor needed to keep the actual current below this value.

LED BJT Drive

Often you need to reduce the current drawn from a GPIO line. The Bipolar Junction Transistor (BJT) may be relatively old technology, but it is a current amplifier, low in cost and easy to use.

A BJT is a three-terminal device - base, emitter and collector - in which the current that flows through the emitter/collector is controlled by the current in the base:

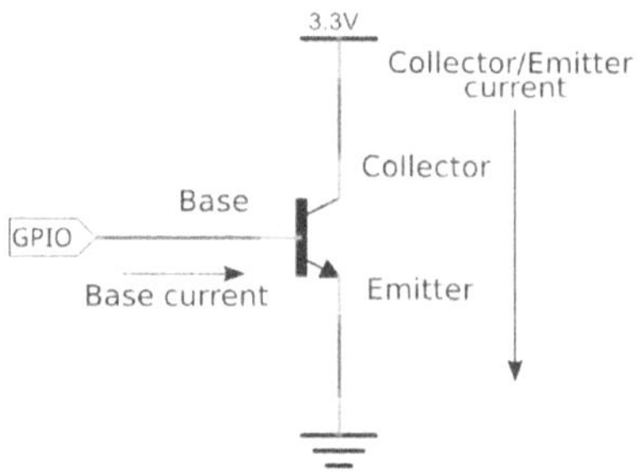

The diagram shows an NPN transistor, which is the most common. This diagram is a simplification in that, in reality, the current in the emitter is slightly larger than that in the collector because you have to add the current flowing in the base.

In most cases you need just two additional facts. Firstly, the voltage on the base is approximately 0.6V, no matter how much current flows since the base is a diode, a non-linear device just like the LED in the previous section. Secondly, the current in the collector/emitter is hfe or ß (beta) times the current in the base. That is, hfe or beta is the current gain of the transistor and you look it up for any transistor you want to use. While you are consulting the datasheets, you also need to check the maximum currents and voltages the device will tolerate. In most cases, the beta is between 100 and 200 and hence you can use a transistor to amplify the GPIO current by at least a factor of 100.

Notice that, for the emitter/collector current to be non-zero, the base has to have a current flowing into it. If the base is connected to ground then the transistor is "cut off", i.e. no current flows. What this means is that when the GPIO line is high the transistor is "on" and current is flowing and when the GPIO line is low the transistor is "off" and no current flows. This high-on/low-off behavior is typical of an NPN transistor.

A PNP transistor works the other way round:

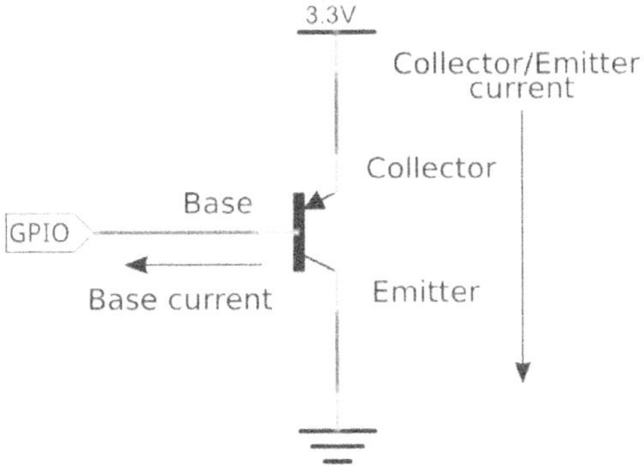

The 0.6V is between the base and the collector and the current flows out of the base. In this case to switch the transistor on you have to connect the base to ground. What this means is that the transistor is off when the GPIO line is high and on when it is low.

This complementary behavior of NPN and PNP BJTs is very useful and means that we can use such transistors in pairs. It is also worth knowing that the diagram given above is usually drawn with 0V at the top of the diagram, i.e. flipped vertically, to make it look the same as the NPN diagram. You always need to make sure you know where the +V line is.

A BJT Example

For a simple example we need to connect a standard LED to a GPIO line with a full 20mA drive. Given that all of the Pi's GPIO lines work at 3.3V and ideally only supply a few milliamps, we need a transistor to drive the LED which typically draws 20mA.

You could use a Field Effect Transistor (FET) of some sort, but for this sort of application an old-fashioned BJT (Bipolar Junction Transistor) works very well and is cheap and available in a thru-hole mount, i.e. it comes with wires. Almost any general purpose NPN transistor will work, but the 2N2222 is very common. From its datasheet you can discover that the max collector current is 800mA and beta is at least 50, which makes it suitable for driving a 20mA LED with a GPIO current of at most 20/50mA = 0.4mA.

The circuit is simple but we need two current-limiting resistors:

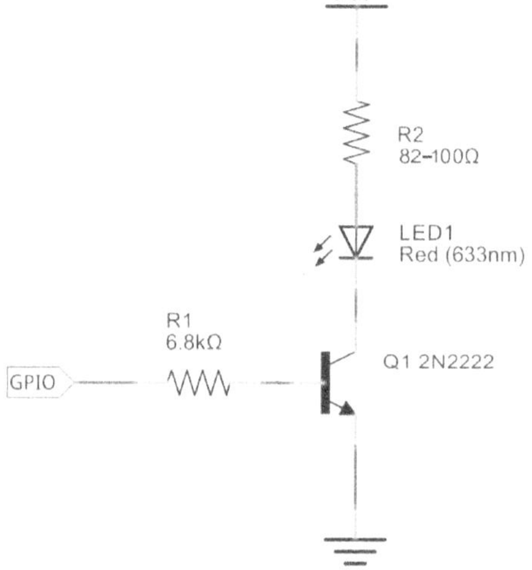

If you connected the base to the GPIO line directly then the current flowing in the base would be unrestricted – it would be similar to connecting the GPIO line to ground. R1 restricts the current to 0.39mA, which is very low and, assuming that the transistor has a minimum gain (hfe) of 50, this provides just short of 20mA to power it.

The calculation is that the GPIO supplies 3.3V and the base has 0.6V across it so the voltage across R1 is 3.3 - 0.6V = 2.7V. To limit the current to 0.4mA would need a resistor of 2.7/0.4kΩ = 6.7kΩ. The closest preferred value is 6.8kΩ, which gives a slightly smaller current.

Without R2 the LED would draw a very large current and burn out. R2 limits the current to 20mA. Assuming a forward voltage drop of 1.6V and a current of 20mA the resistor is given by (3.3-1.6)/20kΩ = 85Ω. In practice, we could use anything in the range 82Ω to 100Ω.

The calculation just given assumes that the voltage between the collector and emitter is zero, but of course in practice it isn't. Ignoring this results in a current less than 20mA, which is erring on the safe side. The datasheet indicates that the collector emitter voltage is less than 200mV.

The point is that you rarely make exact calculations for circuits such as this, you simply arrive at acceptable and safe operating conditions.

You can also use this design to drive something that needs a higher voltage. For example, to drive a 5V dip relay, which needs 10mA to activate it, you would use something like:

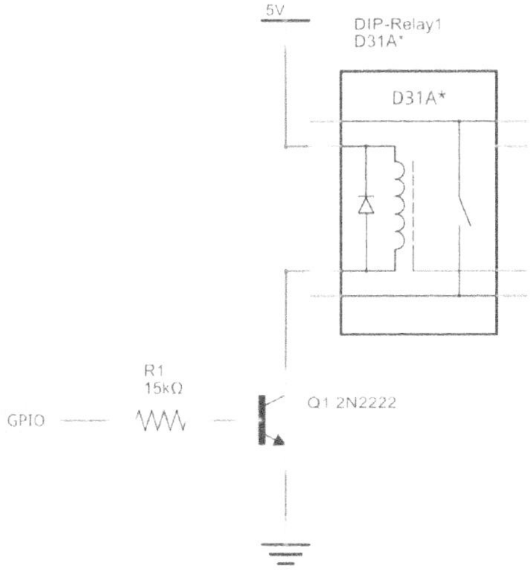

Notice that in this case the transistor isn't needed to increase the drive current – the GPIO line could provide the 10mA directly. Its purpose is to change the voltage from 3.3V to 5V. The same idea works with any larger voltage.

If you are using the 2N2222 then the pinouts are:

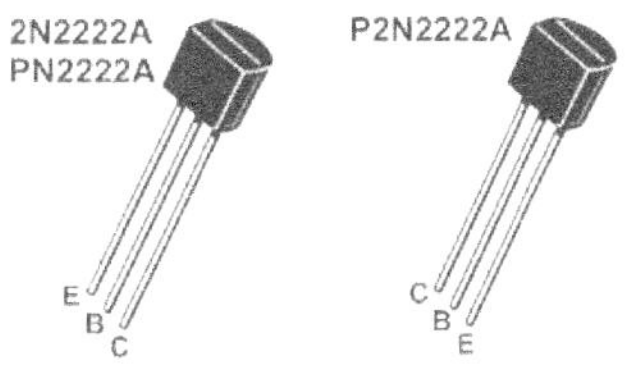

As always, the positive terminal on the LED is the long pin.

MOSFET Driver

There are many who think that the FET (Field Effect Transistor) or more precisely the MOSFET (Metal Oxide Semiconductor FET) is the perfect amplification device and we should ignore BJTs. They are simpler to understand and use, but it can be more difficult to find one with the characteristics you require.

Like the BJT, a MOSFET has three terminals called the gate, drain and source. The current that you want to control flows between the source and drain and it is controlled by the gate. This is analogous to the BJT's base, collector and emitter, but the difference is that it is the voltage on the gate that controls the current between the source and drain.

The gate is essentially a high resistance input and very little current flows in it. This makes it an ideal way to connect a GPIO line to a device that needs more current or a different voltage. When the gate voltage is low the source drain current is very small. When the gate voltage reaches the threshold voltage $V_{GS(th)}$, which is different for different MOSFETs, the source drain current starts to increase exponentially. Basically, when the gate is connected to 0V or below $V_{GS(th)}$ the MOSFET is off and when it is above $V_{GS(th)}$ the MOSFET starts to turn on. Don't think of $V_{GS(th)}$ as the gate voltage that the MOSFET turns on, but as the voltage below which it is turned off.

The problem is that the gate voltage to turn a typical MOSFET fully on is in the region of 10V. Special "logic" MOSFETs need a gate voltage around 5V to fully turn on and this makes the 3.3V at which the Pico's GPIO lines work a problem. The datasheets usually give the fully on resistance and the minimum gate voltage that produces it, usually listed as Drain-Source On-State Resistance. For digital work this is a more important parameter than the gate threshold voltage.

You can deal with this problem in one of two ways – ignore it or find a MOSFET with a very small $V_{GS(th)}$. In practice MOSFETs with thresholds low enough to work at 3.3V are hard to find and when you do find them they are generally only available as surface-mount. Ignoring the problem sometimes works if you can tolerate the MOSFET not being fully on. If the current is kept low then, even though the MOSFET might have a resistance of a few ohms, the power loss and voltage drop may be acceptable.

What MOSFETs are useful for is in connecting higher voltages to a GPIO line used as an input – see later.

Also notice that this discussion has been in terms of an N-channel MOSFET. A P-channel works in the same way, but with all polarities reversed. It is cut off when the gate is at the positive voltage and on when the gate is grounded. This is exactly the same as the NPN versus PNP behavior for the BJT.

MOSFET LED

A BJT is the easiest way to drive an LED, but as an example of using a common MOSFET we can arrange to drive one using a 2N7000, a low-cost, N-channel device available in a standard TO92 form factor suitable for experimentation:

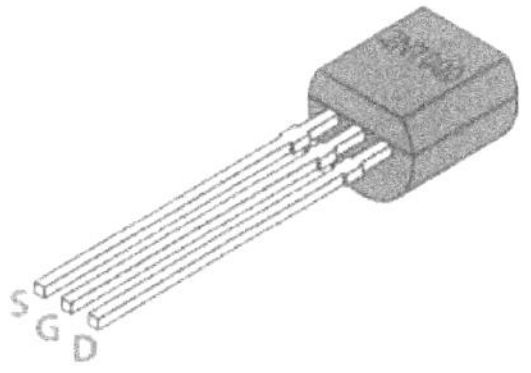

Its datasheet states that it has a $V_{GS(th)}$ typically 2V, but it could be as low as 0.8V or as high as 3V. Given we are trying to work with a gate voltage of 3.3V you can see that in the worst case this is hardly going to work – the device will only just turn on. The best you can do is to buy a batch of 2N7000 and measure their $V_{GS(th)}$ to weed out any that are too high. This said, in practice the circuit given below does generally work.

Assuming a $V_{GS(th)}$ of 2V and a current of 20mA for the LED the datasheet gives a rough value of 6Ω for the on resistance with a gate voltage of 3V. The calculation for the current-limiting resistor is the same as in the BJT case and the final circuit is:

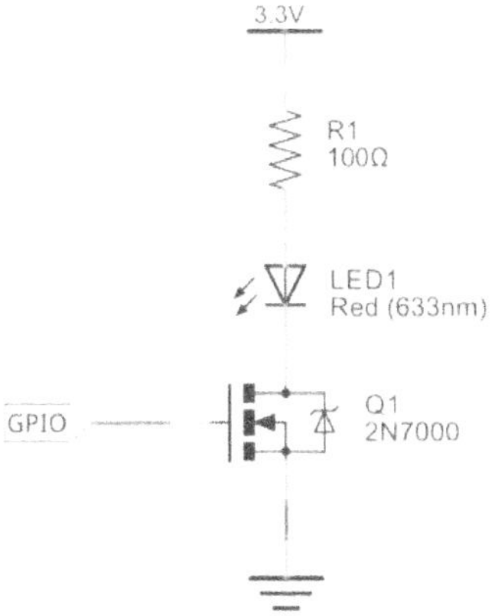

Notice that we don't need a current-limiting resistor for the GPIO line as the gate connection is high impedance and doesn't draw much current. In practice, it is usually a good idea to include a current-limiting resistor in the GPIO line if you plan to switch it on and off rapidly. The problem is that the gate looks like a capacitor and fast changes in voltage can produce high currents. Notice that there are likely to be devices labeled 2N7000 that will not work in this circuit due to the threshold gate voltage being too high, but encountering one is rare.

A logic-level MOSFET like the IRLZ44 has a resistance of 0.028Ω at 5V compared to the 2N2222's of 6Ω. It also has a $V_{GS(th)}$ guaranteed to be between 1V and 2V. It would therefore be a better candidate for this circuit.

Setting Drive Type

The GPIO output can be configured into one of a number of modes, but the most important is pull-up/down. Before we get to the code to do the job it is worth spending a moment explaining the three basic output modes, push-pull, pull-up and pull-down.

Push-Pull Mode

In push-pull mode two transistors of opposite polarity, one PNP and one NPN, are used:

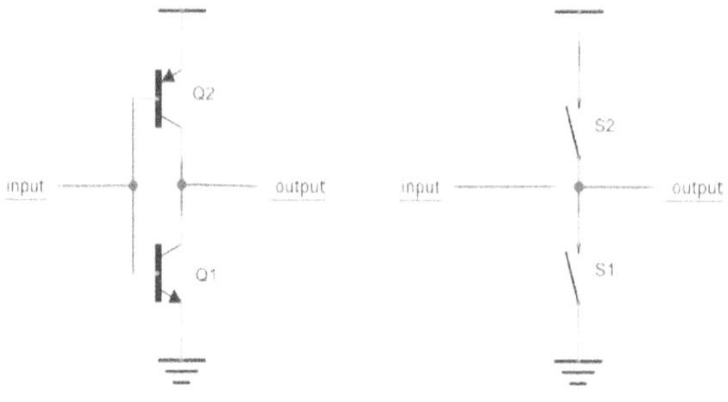

The circuit behaves like the two-switch equivalent shown on the right. Only one of the transistors, or switches, is "closed" at any time. If the input is high then Q1 is saturated and the output is connected to ground - exactly as if S1 was closed. If the input is low then Q2 is saturated and it is as if S2 was closed and the output is connected to 3.3V. You can see that this pushes the output line high with the same "force" as it pulls it low. This is the standard configuration for a GPIO output.

Pull-Up Mode

In pull-up mode one of the transistors is replaced by a resistor:

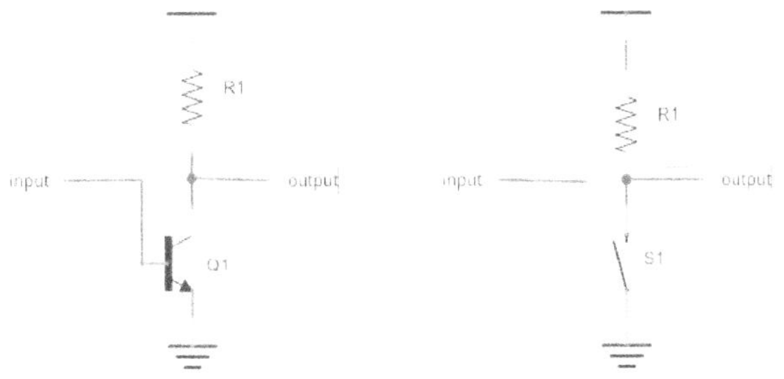

In this case the circuit is equivalent to having a single switch. When the switch is closed, the output line is connected to ground and hence driven low. When the switch is open, the output line is pulled high by the resistor.

You can see that in this case the degree of pull-down is greater than the pull-up, where the current is limited by the resistor. The advantage of this mode is that it can be used in an AND configuration. If multiple GPIO or other lines are connected to the output, then any one of them being low will pull the output line low. Only when all of them are off does the resistor succeed in pulling the line high. This is used, for example, in a serial bus configuration like the I2C bus.

Pull-Down Mode

Finally the pull-down mode, which is the best mode for driving general loads, motors, LEDs, etc, is exactly the same as the pull-up only now the resistor is used to pull the output line low.

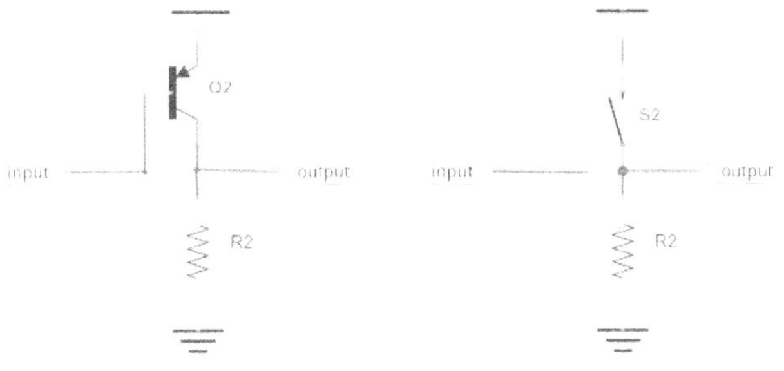

The line is held high by the transistor and pulled low by the resistor only when all the switches are open. Putting this the other way round, the line is high if any one switch is closed. This is the OR version of the shared bus idea.

Setting Output Mode

MicroPython's `Pin` class has some additional parameters in the constructor to set the mode for a GPIO line:

```
machine.Pin(id, mode, pull, value)
```

We have already met `id` and `mode` and the `pull` parameter can be any of:

`None` - No pull-up/down resistor

`Pin.PULL_UP = 1` - Pull-up resistor enabled.
`Pin.PULL_DOWN = 2` - Pull-down resistor enabled.

In addition to these you can also set `mode` to `Pin.OPEN_DRAIN = 2`, which gives you an active low and a high impedance for the `1` state.

If you set the line to output, i.e. `OUT` or `OPEN_DRAIN` then you can set an initial state using `value`.

After you have created a `Pin` object you can change its configuration using the `init` method which takes the same parameters:

```
Pin.init(mode, pull,  value)
```

Only the parameters specified are changed.

The `init` method is particular useful for changing a line from input to output and vice versa.

Basic Input Circuit - The Switch

Now it is time to turn our attention to the electrical characteristics of GPIO lines as inputs. One of the most common input circuits is the switch or button.

Many beginners make the mistake of wiring a GPIO line to a switch something like:

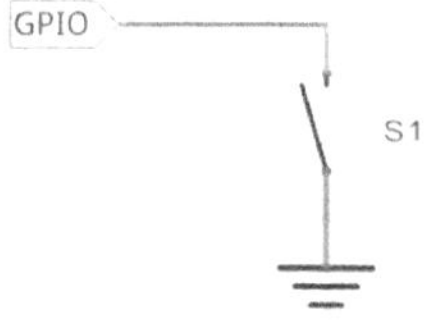

The problem with this is that, if the switch is pressed, the GPIO line is connected to ground and will read as zero. The question is, what does it read when the switch is open? A GPIO line configured as an input without pull up or pull down enabled has a very high resistance. It isn't connected to any particular voltage and the voltage on it varies due to the static it picks up. The jargon is that the unconnected line is "floating". When the switch is open the line is floating and if you read it the result, zero or one, depends on whatever noise it has picked up.

The correct way to do the job is to tie the input line either high or low when the switch is open using a resistor. A pull-up arrangement would be something like:

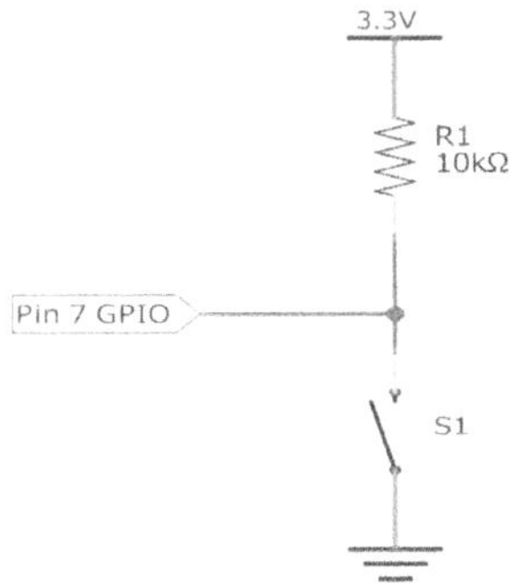

The value of the resistor used isn't critical. It simply pulls the GPIO line high when the switch isn't pressed. When it is pressed a current of a little more than 0.3mA flows in the resistor. If this is too much, increase the resistance to 100kΩ or even more - but notice that the higher the resistor value the noisier the input to the GPIO and the more it is susceptible to RF interference. Notice that this gives a zero when the switch is pressed.

If you want a switch that pulls the line high instead of low, reverse the logic by swapping the positions of the resistor and the switch in the diagram to create a pull-down:

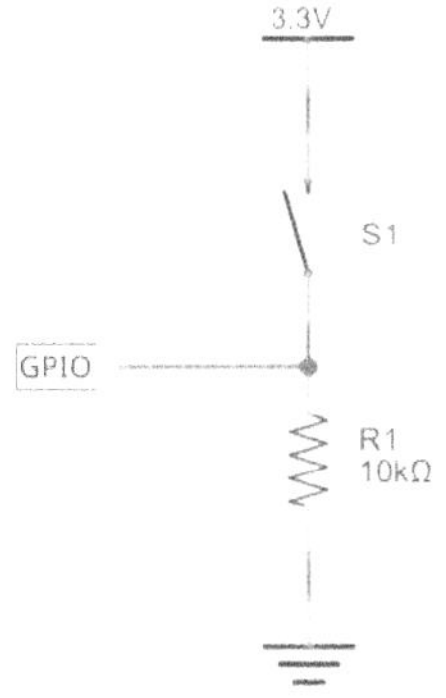

Notice that this gives a one when the switch is pressed.

The good news is that the Pico defaults to an input configuration with a pull-up resistor of around 50kΩ which means you can connect a switch directly to a default GPIO line and it will give a zero when the switch is pressed.

Debounce

Although the switch is the simplest input device, it is very difficult to get right. When a user clicks a switch of any sort, the action isn't clean - the switch bounces. What this means is that the logic level on the GPIO line goes high then low and high again and bounces between the two until it settles down. There are electronic ways of debouncing switches, but software does the job much better. All you have to do is insert a delay of a millisecond or so after detecting a switch press and read the line again - if it is still low then record a switch press. Similarly, when the switch is released, read the state twice with a delay. You can vary the delay to modify the perceived characteristics of the switch.

A more sophisticated algorithm is based on the idea of integration to debounce a switch. All you have to do is read the state multiple times, every few milliseconds say, and keep a running sum of values. If you sum say ten values each time then a total of between 6 and 10 can be taken as an indication that the switch is high. A total less than this indicates that the switch is low. You can think of this as a majority vote in the time period for the switch being high or low.

The Potential Divider

If you have an input that is outside of the range of 0V to 3.3V then you can reduce it using a simple potential divider. In the diagram `V` is the input from the external logic and `Vout` is the connection to the GPIO input line:

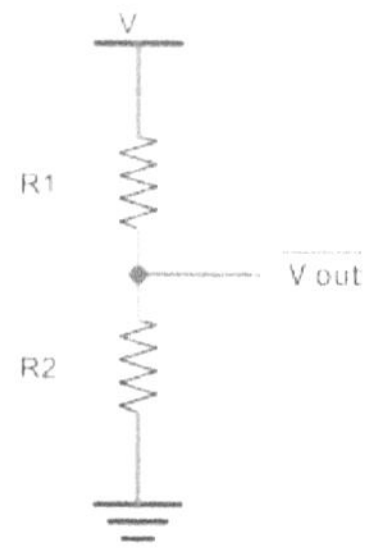


```
Vout = V R2/(R1+R2)
```

You can spend a lot of time on working out good values of `R1` and `R2`. For loads that take a lot of current you need `R1+R2` to be small and divided in the same ratio as the voltages.

For example, for a 5V device R1=18KΩ or 20KΩ and R2=33KΩ work well to drop the voltage to 3.3V.

A simpler approach that works for a 5V signal is to notice that the ratio R1:R2 has to be the same as (5-3.3):3.3, i.e. the voltage divides itself across the resistors in proportion to their value, which is roughly 1:2. What this means is that you can take any resistor and use it for R1 and use two of the same value in series for R2 and the `Vout` will be 3.3V.

The problem with a resistive divider is that it can round off fast pulses due to the small capacitive effects. This usually isn't a problem, but if it is then the solution is to use a FET or a BJT as an active buffer:

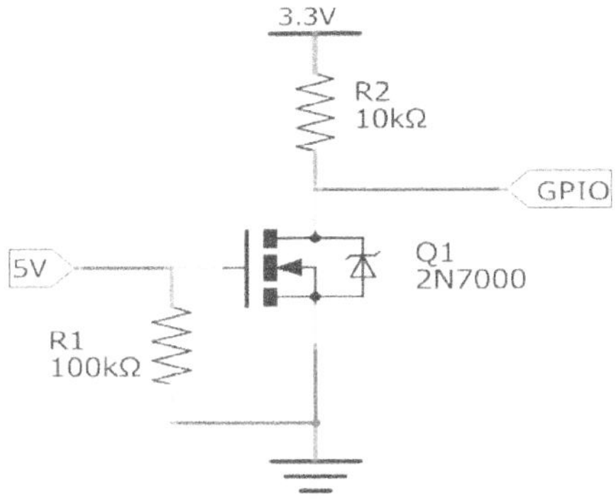

Notice that this is an inverting buffer, the output is low when the input is high, but you can usually ignore this and simply correct it in software, i.e. read a `1` as a low state and a `0` as a high state. The role of R1 is to make sure the FET is off when the 5V signal is absent and R2 limits the current in the FET to about 0.3mA. In most cases you should try the simple voltage divider and only move to an active buffer if it doesn't work.

This very basic look at electronics isn't all that you need to know, but it is enough for you to see some of the problems and find some answers. In general, this sort of electronics is all about making sure that voltages and currents are within limits. As switching speeds increase you have additional problems, which are mainly concerned with making sure that your circuits aren't slowing things down. This is where things get more subtle.

Summary

- You can get a long way with only a small understanding of electronics, but you do need to know enough to protect the Pico and things you connect to it.
- The maximum current from any GPIO line should be less than 12mA and the total current should be less than 30mA.
- All of the GPIO lines work at 3.3V and you should avoid directly connecting any other voltage.
- You can drive an LED directly from a GPIO line, but only at 16mA rather than the nominal 20mA needed for full brightness.
- Calculating a current-limiting resistor always follows the same steps – find out the current in the device, find out the voltage across the device and work out the resistor that supplies that current when the remainder of the voltage is applied to it.
- For any load you connect to a GPIO output, you generally need a current-limiting resistor.
- In many cases you need a transistor, a BJT, to increase the current supplied by the GPIO line.
- To use a BJT you need a current-limiting resistor in the base and generally one in the collector.
- MOSFETs are popular alternatives to BJTs, but it is difficult to find a MOSFET that works reliably at 3.3V.
- GPIO output lines can be set to active push-pull mode, where a transistor is used to pull the line high or low, or passive pull-up or pull-down mode, where one transistor is used and a resistor pulls the line high or low when the transistor is inactive.
- GPIO lines have built-in pull-up and pull-down resistors which can be selected or disabled under software control.
- When used as inputs, GPIO lines have a very high resistance and in most cases you need pull-up or pull-down resistors to stop the line floating.
- The built-in pull-up or pull-down resistors can be used in input mode.
- Mechanical input devices have to be debounced to stop spurious input.
- If you need to connect an input to something bigger than 3.3V then you need a potential divider to reduce the voltage back to 3.3V. You can also use a transistor.

Chapter 6
Simple Input

There is no doubt that input is more difficult than output. When you need to drive a line high or low you are in command of when it happens, but input is in the hands of the outside world. If your program isn't ready to read the input, or if it reads it at the wrong time, then things just don't work. What is worse, you have no idea what your program is doing relative to the event you are trying to capture. Welcome to the world of input.
In this chapter we look at the simplest approach to input – the polling loop. This may be simple, but it is a good way to approach many tasks. In Chapter 8 we look at more sophisticated alternatives – events and interrupts.

GPIO Input

GPIO input is a much more difficult problem than output from the point of view of measurement and verification. For output at least you can see the change in the signal on a logic analyzer and know the exact time that it occurred. This makes it possible to track down timing problems and fine tune things with good accuracy.

Input on the other hand is "silent" and unobservable. When did you read in the status of the line? Usually the timing of the read is with respect to some other action that the device has taken. For example, you read the input line 20μs after setting the output line high. But how do you know when the input line changed state during that 20 microseconds? The simple answer is in most cases you don't.

In some applications the times are long and/or unimportant but in some they are critical and so we need some strategies for monitoring and controlling read events. The usual rule of thumb is to assume that it takes as long to read a GPIO line as it does to set it. This means we can use the delay mechanisms that we looked at with regard to output for input as well.

One common and very useful trick when you are trying to get the timing of input correct is to substitute an output command to a spare GPIO line and monitor it with a logic analyzer. Place the output instruction just before the input instruction and where you see the line change on the logic analyzer should be close to the time that the input would be read in the unmodified program. You can use this to debug and fine tune and then remove the output statement.

Basic Input Functions

The `Pin` object can be set to input mode using the constructor:

```
pin = Pin(22, Pin.IN)
```

You can also set the direction to input using the `init` method:

```
pin.init(mode=Pin.IN)
```

Once set to input, the GPIO line is high impedance so it won't take very much current, no matter what you connect it to. However, notice that the Pico uses 3.3V logic and you should not exceed this value on an input line. For a full discussion of how to work with input see the previous chapter.

You can read the line's input state using the `value` method:

```
result=pin.value()
```

Notice that this is an "overloaded" method. If you supply a value as a parameter then it attempts to set the value as output. If you don't specify a value then it gets the GPIO level as a zero or a one.

This is all there is to using a GPIO line as an input, apart from the details of the electronics and the small matter of interrupts.

As introduced in the previous chapter you can also set the internal pull-up or pull-down resistors using one of:

`Pin.PULL_UP = 1`	Pull-up resistor enabled
`Pin.PULL_DOWN = 2`	Pull-down resistor enabled

in the constructor or the `init` method.

The pull-up/down resistors are between 50 and 80kΩ.

The Simple Button

One of the most common input circuits is the switch or button. If you want another external button you can use any GPIO line and the circuit explained in the previous chapter. That is, the switch has to have either a pull-up or pull-down resistor either provided by you or a built-in one enabled using software.

The simplest switch input program using an internal pull-up is:

```
from machine import Pin
import time
pinIn = Pin(22, Pin.IN,Pin.PULL_DOWN)
pinLED = Pin(25, Pin.OUT)

while True:
    if pinIn.value():
        pinLED.on()
    else:
        pinLED.off()
    time.sleep(0.5)
```

As the internal pull-down resistor is used, the switch can be connected to the line and ground without any external resistors:

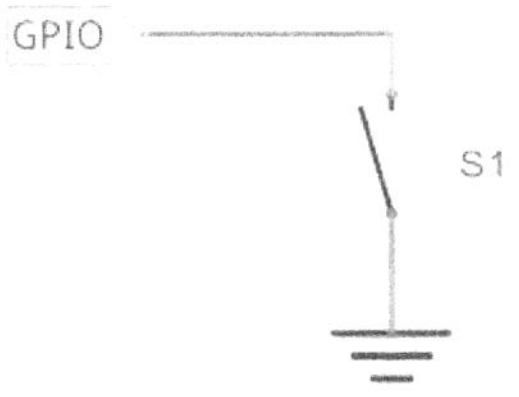

The program simply tests for the line to be pulled high by the switch not being closed and then sets GP25 high. As GP25 is connected the on-board LED it will light up while it is not pressed. Notice GP22 goes low to indicate that the switch is pressed.

If you change `PULL_DOWN` to `PULL_UP`, the way the switch is connected becomes:

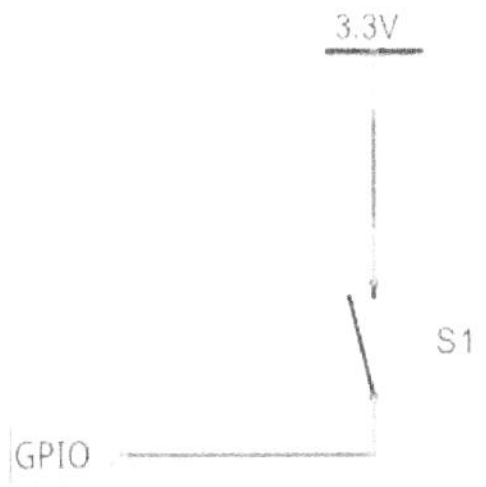

The program still works, but now GP22 is high when the switch is pressed and hence the LED is on when the switch is pressed.

Should you use an internal or external resistor? The answer is that it mostly doesn't matter as long as there is a resistor. The only problem with using an internal resistor is the possibility that the software fails to set the pull-up/down mode and leaves the input floating.

Also notice that this switch input is not debounced. The simplest way to do this is include a time delay in the loop before the line is sampled again.

If you want to respond to a button press, that is a press and a release event, then you have to test for a double transition:

```
from machine import Pin
import time
pinIn = Pin(22, Pin.IN,Pin.PULL_DOWN)
pinLED = Pin(25, Pin.OUT)

while True:
    while pinIn.value()==0:
        pass
    while pinIn.value()==1:
        pass
    pinLED.on()
    time.sleep(1)
    pinLED.off()
```

In this case you really do need the debounce delays if you want to avoid responding twice to what the user perceives as a single press.

A 1-millisecond delay is probably the smallest delay that produces a button that feels as if it works. In practice, you would have to tune the delay to suit the button mechanism in use and the number of times you can allow the button to be pressed in one second.

Press Or Hold

You can carry on elaborating on how to respond to a button. For example, most users have grown accustomed to the idea that holding a button down for a longer time than a press makes something different happen.

To distinguish between a press and a hold all you need to do is time the difference between line down and line up:

```
from machine import Pin
import time
pinIn = Pin(22, Pin.IN,Pin.PULL_DOWN)
pinLED = Pin(25, Pin.OUT)

while True:
    while pinIn.value()==0:
        pass
    t=time.ticks_ms()
    time.sleep_ms(1)
    while pinIn.value()==1:
        pass
    t=time.ticks_diff(time.ticks_ms(),t)
    if t<2000:
        pinLED.on()
        time.sleep(1)
        pinLED.off()
    else:
        for i in range(10):
            pinLED.on()
            time.sleep_ms(100)
            pinLED.off()
            time.sleep_ms(100)
```

In this case holding the button for 2s registers a "held" – the LED flashes 10 times and anything less is a "push" – the LED flashes just once. Notice the 1ms debounce pause between the test for no-press and press.

One of the problems with all of these sample programs is that they wait for the button to be pressed or held and this makes it difficult for them to do anything else. You have to include whatever else your program needs to do within the loop that waits for the button – the polling loop. You can do this in an ad hoc way, but the best approach is to implement a finite state machine, see later.

How Fast Can We Measure?

Buttons are one form of input, but often we want to read data from a GPIO line driven by an electronic device and decode the data. This usually means measuring the width of the pulses and this raises the question of how fast can we accept input?

The simplest way to find out how quickly we can take a measurement is to perform a pulse width measurement. Applying a square wave to GP22 we can measure the time that the pulse is high using:

```
from machine import Pin
import time
pinIn = Pin(22, Pin.IN)

while True:
    while pinIn.value()==1:
        pass
    while pinIn.value()==0:
        pass

    t=time.ticks_us()
    while pinIn.value()==1:
        pass
    t=time.ticks_diff(time.ticks_us(),t)
    print(t)
    time.sleep(1)
```

This might look a little strange at first. The inner `while` loops are responsible for getting us to the right point in the waveform. First we loop until the line goes low, then we loop until it goes high again and finally measure how long before it goes low. You might think that we simply have to wait for it to go high and then measure how long till it goes low, but this misses the possibility that the signal might be part way through a high period when we first measure it. This can be measured down to around $30\mu s$ with a very poor accuracy at this limit and an accuracy of 5 to $10\mu s$ for longer times .

Notice that in either case if you try measuring pulse widths much shorter than the lower limit that works, you will get results that look like longer pulses are being applied. The reason is simply that the Pico will miss the first transition to zero but will detect a second or third or later transition. This is the digital equivalent of the aliasing effect found in the Fourier Transform or general signal processing.

MicroPython also provides a special method that implements the algorithm used in the above program. The method:

```
machine.time_pulse_us(pin, level, timeout)
```

will block until the pin specified changes to the level specified or the timeout occurs. Once the pin attains the level specified it times how long it takes for it to change and returns the result in microseconds. The only problem is that, if the pin is already at the specified level, the timing begins at once.

Using this method the program above can be written more simply as:

```
import time
import machine

pinIn = machine.Pin(22, machine.Pin.IN)

while True:
    t = machine.time_pulse_us(pinIn, 1)
    print(t)
    time.sleep(1)
```

Notice that you will get some fractional measurements using this method. To get accurate results you would have to be sure that the line was at zero just before the method is used. For example:

```
import machine
import time

import machine
pinIn = machine.Pin(22, machine.Pin.IN)

while True:
    while pinIn.value() == 1:
        pass
    t = machine.time_pulse_us(pinIn, 1)
    print(t)
    time.sleep(1)
```

The `time_pulse_us` method is not a big improvement on the more basic way of doing the job, but it is more accurate for lengthier intervals.

The Finite State Machine

If your project requires a complex set of input and output lines then you need an organizing principle to save you from the complexity. When you first start writing IoT programs that respond to the outside world you quickly discover that all of your programs take a similar form:

```
while(True):
  wait for some input lines
  process the input data
  write some output lines
  wait for some input lines
  read some more input lines
  write some output lines
```

For most programmers this is a slightly disturbing discovery because programs are not supposed to consist of infinite loops, but IoT programs nearly always, in principle if not in practice, take the form of an apparently infinite polling loop. A second, and more important, aspect is that the way in which reading and writing GPIO lines is related can be so complex that it can be difficult to see exactly when any particular line is read and when it is written.

It is natural to try to find implementations that make this simpler. In most cases, programmers discover or invent the idea of the event or, better, the interrupt. In this case when something happens in the outside world a function is automatically called to deal with it and the relation between the external state and the system's response is seemingly well-defined. Of course, in practice it isn't, as you have to deal with what happens when multiple events or interrupts occur at the same, or very nearly the same, time.

Often more sophisticated approaches are used to try and handle more external changes of state in a given time. Somehow the infinite polling loop is seen as wasteful. What is the CPU doing if it spends all its time looping round waiting for something to happen? Of course, if it has nothing better to do then it isn't a waste. In fact IoT devices are often dedicated to just getting one job done so the "polling is wasteful" meme, so prevalent in the rest of computing, is completely unjustified.

What is more, the polling loop is usually the way to get the greatest throughput. If a processor can handle X external state changes per second and respond to these with Y external state changes per second, then moving to an event- or interrupt-based implementation reduces both X and Y. In short, if a processor cannot do the job using a polling loop then it cannot do the job.

This is not to say that there aren't advantages to events and interrupts – there are, but they don't increase throughput.

So how should you organize a polling loop so that what it does is self-evident by looking at the code?

There are many answers to this according to the system being implemented and there are no "pure" theoretical answers that solve all problems, but the finite state machine, or FSM, is a model every IoT programmer should know.

A finite state machine is very simple. At any given time the machine/program has a record of the current state S. At a regular interval the external world provides an input I which changes the state from S to S' and produces an output O. That's all there is to a finite state machine. There are variations on the definition of the FSM but this one, a Mealy machine because its outputs depend on both its state and the input, is most suitable for IoT programming.

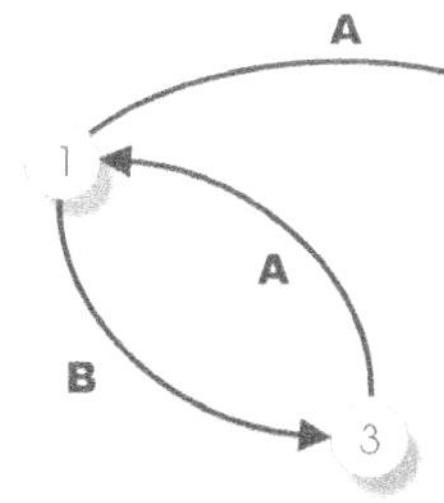

You can design an FSM with the help of a diagram. In the FSM shown we have three states 1, 2 and 3 and if you are in state 1 an input of A moves the system to state 2 and an input of B moves it into state 3.

Your program simply needs to take the form of a polling loop that implements a finite state machine. It reads the input lines as I and uses this and the current state S to determine the new state S' and the output O. There is some overhead in using this organization, but it is usually worth it. Notice that this organization implies that you read input once, make changes once and set outputs once in the loop. If you fix the time that the polling loop takes then you know the characteristic time for any changes to the system.

FSM Button

As an example, let's implement the simple button given earlier in the chapter. The first version used multiple loops to wait for changes in the state of the input line. The finite state version uses only a single polling loop:

```
from machine import Pin
import time
pinIn = Pin(22, Pin.IN)

s=0
count=0
while True:
    i=pinIn.value()
    t=time.ticks_add(time.ticks_us(),1000*100)
    if s==0:   #button not pushed
        if i:
            s=1
            count=count+1
            print("Button Push ",count)
    elif s==1: #button pushed
        if not i:
            s=0
    else:
        s=0
    while time.ticks_us()<t:
        pass
```

This looks more complicated than the original, and there are more lines of code, but it is much easier to extend to more complex situations. In this case we have only two states – `s = 0` for button not pushed and `s = 1` for button pushed. Ideally the states that we use shouldn't refer to inputs or outputs, but to the overall state of the system. For example, if you were using a Pico to control a nuclear reactor you might use a state "CoreMeltdown" in preference to "TempSensorOverLimit". States should be about the consequence of the inputs and the outputs should be the consequence of the current state. In the example above the inputs and output are too simple to give rise to an abstract concept of "state". Even if you were to change the state labels to "LEDOn" or "LEDOff" they are directly related to the state of a single output line.

The key idea, however, is that the states indicate the state of the system at the time of the input. That is, `s = 0` (button not pushed) is the state when the system reads in a low on the GPIO line (recall the line is pulled low so pushing the button makes it go high). You can see at the start of the polling loop we read the input line and store its value in the variable `i`. Next, a case statement is used to process the input depending on the current state. You can see that if `s = 0`, i.e. button not pushed, then the state moves to `s = 1`, i.e. button pushed, and a message is printed giving the number of times the button has been pressed as a simple action. In general, the action could be setting a GPIO line high or doing anything that is appropriate for the new state. Notice that actions occur on state changes.

If the state is in `s = 1`, i.e. button pushed, then the input has to be `0` for anything to happen. In this case the state changes to `s = 0` and any actions that are needed to take the system from state `1` to state `0` are performed – none in this case. Finally, if the state is anything other than `0` or `1`, we set it to `0` as something is wrong.

Notice that the polling loop is set up so that the whole thing repeats every 100ms. The time is taken at the start of the loop and after everything has been processed we wait for 100ms to be up. What this means is that, no matter how long the processing in the loop takes, as long as it takes less than 100ms the loop will repeat every 100ms.

This is a very simple finite state machine polling loop. In practice, there is usually a set of ifs that deals with each current state, but there is often another set of if statements within each state case to deal with what happens according to different inputs. An alternative way of designing a finite state machine loop is to use a lookup table, indexed by state and input, which gives you the new state and the actions.

FSM Hold Button

As a slightly more complicated example of using the FSM approach, let's implement the button with hold. You might think that button with hold has three states – button not pushed, button pushed and button held. You can implement it in this way, but there is an argument that there are still only two states – not pushed and pushed. The held state is better implemented as extra input data to the state, i.e. the time the button has been in the pressed state.

Remember, the output of a FSM depends on the state and the input and in this case the input is the line level and the time pressed:

```
from machine import Pin
import time
pinIn = Pin(22, Pin.IN)

s=0
while True:
    i=pinIn.value()
    t=time.ticks_add(time.ticks_us(),1000*100)
    if s==0:        #button not pushed
        if i:
            s=1
        tpush=t
    elif s==1:      #button pushed
        if not i:
            s=0
            if time.ticks_diff(t, tpush) > 2000000:
                print("Button held \n\r")
            else:
                print("Button pushed \n\r")
    else:
        s=0
    while time.ticks_us()<t:
        pass
```

It is clear that you can't know the time the button has been pressed until it is released so the actions are now all in the button-pushed state. While the button is in the pushed state it can be released and we can compute the time it has been pressed and modify the action accordingly.

FSM Ring Counter

Another very common input configuration is the ring counter. A ring counter moves on to a new output each time it receives an input and repeats when it reaches the last output of the set. For example, if you have three output lines connected to three LEDs then initially LED 0 is on, when the user presses the button LED 1 is on and the rest off, the next user press moves on to LED 2 on and another press turns LED 0 on. You can see that as the user keeps pressing the button the LEDs go on and off in a repeating sequence.

A common implementation of a ring counter has a state for each button press and release for each LED being on. For three LEDs this means six states and this has a number of disadvantages. A better idea is to have just two states, button pressed and button released and use a press counter as an additional input value. This means that what happens when you enter the button pressed state depends on the value in the counter.

We also change from using the measurement of the button as pressed or released and move to considering an "edge" signal. Generally we need inputs that indicate an event localized in time. Button "pressed" and button "released" are events that are extended in time but "press" and "release" are localized to small time intervals that can be thought of as single time measurements. In general we prefer "edge" signals because these indicate when something has changed.

Implementing this is fairly easy:

```
from machine import Pin
import time
pinIn = Pin(22, Pin.IN)
pinLED1 = Pin(21, Pin.OUT)
pinLED2 = Pin(20, Pin.OUT)
pinLED3 = Pin(19, Pin.OUT)
pinLED1.on()
pinLED2.off()
pinLED3.off()
s=0
buttonState=pinIn.value()
while True:
    buttonNow=pinIn.value()
    edge=buttonState-buttonNow
    buttonState=buttonNow
    t=time.ticks_add(time.ticks_us(),1000*100)
    if s==0:
        if edge:
            s=1
            pinLED1.off()
            pinLED2.on()
            pinLED3.off()

    elif s==1:
        if edge:
            s=2
            pinLED1.off()
            pinLED2.off()
            pinLED3.on()
    elif s==2:
        if edge:
            s=0
            pinLED1.on()
            pinLED2.off()
            pinLED3.off()
    else:
        s=0
    while time.ticks_us()<t:
        pass
```

First we set up the GPIO lines for input and output and set the outputs so that LED `0` is on, i.e. `s = 0`. Next we start the polling loop. Inside the loop there is a `switch` statement that manages three states. At the start of the loop the difference between the current button value and its previous value are used to calculate `edge` which is `1` only when the button has changed from pressed, `1`, to released, `0`. That is, `edge=1` only on a down-going edge. If the button has just been pressed then the state is moved on to the next state, `0→1`, `1→2` and `2→0`, and the LEDs are set to the appropriate values.

You might wonder why all three LEDs are set and not just the two that are changing? There are a number of reasons including it's easier to see what is happening from the code and it makes sure that all of the LEDs are in the state you intend. Notice that polling loop is set up to repeat every 100ms so providing debouncing and a predictable service time. If you try this out you will find that the LEDs light up sequentially on each button press.

Like many more advanced methods the FSM approach can make things seem more complicated in simple examples, but it repays the effort as soon as things get more complicated. A polling loop with tens of states and lots of input and outline lines to manage becomes impossible to maintain without some organizing principle.

Summary

- Input is hard because things happen at any time, irrespective of what your program might be doing.

- You can call the `value` method at any time to discover the state of a GPIO line – the problem is when and how often to call it.

- You can chose between external or internal pull-up/down resistors.

- Mechanical input devices such as buttons have to be debounced.

- The power of software is that it can enhance a simple device. A simple button is either pushed or released, but you can use this to generate a third "held" state.

- Using a polling loop you can handle inputs as short as a few tens of microseconds.

- Most IoT programs are best written as a polling loop.

- The Finite State Machine (FSM) is one way of organizing a complex polling loop so that inputs are tested and outputs are set once for each time through the loop.

- Ideally the states of a FSM should not be simple statements of the inputs of outputs that determine the state, but for simple systems this can be difficult to achieve.

- It can be difficult to work out what constitutes the events of a FSM. Ideally they should be localized in time so that they indicate the moment that something happens.

Chapter 7

Advanced Input - Events and Interrupts

When you start to work with multiple inputs that mean a range of different things, input really becomes a challenge. You can control much of the complexity using finite state machines and similar organizational principles, but sooner or later you are going to have to deal with the problem of input when your program isn't ready for it. There are two general and closely related ways to deal with this problem – events and interrupts. The Pico SDK and MicroPython don't really support events, preferring interrupts, but its hardware certainly is event-capable and it isn't difficult to add some software to make it work. In this chapter we look first at events and then at interrupts.

Events

An event is like a latch or a memory that something happened. Imagine that there is a flag that will be automatically set when an input line changes state. The flag is set without the involvement of software, or at least any software that you have control over. It is useful to imagine an entirely hardware-based setting of the flag, even if this is not always the case. With the help of an event you can avoid missing an input because the polling loop was busy doing something else. Now the polling loop reads the flag rather than the actual state of the input line and hence it can detect if the line has changed since it last polled. The polling loop resets the event flag and processes the event. Of course, it can't always know exactly when the event happened, but at least it hasn't missed it altogether.

A simple event can avoid the loss of a single input, but what if there is more than one input while the polling loop is unavailable? The most common solution is to create an event queue – that is, a FIFO (first in, first out) queue of events as they occur. The polling loop now reads the event at the front of the queue, processes it and reads the next. It continues like this until the queue is empty, when it simply waits for an event. As long as the queue is big enough, an event queue means you don't miss any input, but input events are not necessarily processed close to the time that they occurred. They should be processed in order, but unless the events are timestamped the program has no idea when they happened.

An event queue is a common architecture, but to work or have any advantages, it needs either multiple cores so that events can always be added to the queue before another occurs or it needs the help of hardware, usually in the form of interrupts. Notice that an event, or an event queue, cannot increase the program's throughput or its latency – the time to react to an input. In fact, an event queue decreases throughput and increases latency due to overheads of implementation. All an event system does is to ensure that you do not miss any input and that all input gets processed eventually.

Interrupts Considered Harmful?

Interrupts are often confused with events, but they are very different. An interrupt is a hardware mechanism that stops the computer doing whatever it is currently doing and makes it transfer its attention to running an interrupt handler. You can think of an interrupt as an event flag that, when set, interrupts the current program to run the assigned interrupt handler.

Using interrupts means the outside world decides when the computer should pay attention to input and there is no need for a polling loop. Most hardware people think that interrupts are the solution to everything and polling is inelegant and only to be used when you can't use an interrupt. This is far from the reality. There is a general feeling that real-time programming and interrupts go together and if you are not using an interrupt you are probably doing something wrong. In fact, the truth is that if you are using an interrupt you are probably doing something wrong. So much so that some organizations are convinced that interrupts are so dangerous that they are banned from being used at all.

Interrupts are only really useful when you have a low-frequency condition that needs to be dealt with on a high-priority basis. Interrupts can simplify the logic of your program, but rarely does using an interrupt speed things up because the overhead involved in interrupt handling is usually quite high.

If you have a polling loop that takes 100ms to poll all inputs and there is an input that demands attention in under 60ms then clearly the polling loop is not going to be good enough. Using an interrupt allows the high-priority event to interrupt the polling loop and be processed in less than 100ms. However, if this happens very often the polling loop will cease to work as intended. An alternative is to simply make the polling loop check the input twice per loop.

For a more real-world example, suppose you want to react to a doorbell push button. You could write a polling loop that simply checks the button status repeatedly and forever or you could write an interrupt service routine (ISR) to respond to the doorbell. The processor would be free to get on with other things until the doorbell was pushed when it would stop what it was doing and transfer its attention to the ISR.

How good a design this is depends on how much the doorbell has to interact with the rest of the program and how many doorbell pushes you are expecting. It takes time to respond to the doorbell push and then the ISR has to run to completion - what is going to happen if another doorbell push happens while the first push is still being processed? Some processors have provision for forming a queue of interrupts, but that doesn't help with the fact that the process can only handle one interrupt at a time. Of course, the same is true of a polling loop, but if you can't handle the throughput of events with a polling loop, you can't handle it using an interrupt either, because interrupts add the time to transfer to the ISR and back again.

Finally, before you dismiss the idea of having a processor do nothing but ask repeatedly "is the doorbell pressed", consider what else it has to do. If the answer is "not much" then a polling loop might well be your simplest option. Also, if the processor has multiple cores, then the fastest way of dealing with any external event is to use one of the cores in a fast polling loop. This can be considered to be a software emulation of a hardware interrupt – not to be confused with a software interrupt or trap, which is a hardware interrupt triggered by software.

If you are going to use interrupts to service input then a good design is to use the interrupt handler to feed an event queue. This at least lowers the chance that input will be missed.
Despite their attraction, interrupts are usually a poor choice for anything other than low-frequency events that need to be dealt with quickly.

Hardware Events

MicroPython doesn't support events, but the hardware does as part of its implementation of GPIO interrupts. Each GPIO line records four events:

```
GPIO_IRQ_LEVEL_LOW
GPIO_IRQ_LEVEL_HIGH
GPIO_IRQ_EDGE_FALL
GPIO_IRQ_EDGE_RISE
```

The level events are not latched and simply reflect the current state of the GPIO line. The edge events are latched and stay set until you clear them. In practice, the edge events are far more useful.

Each of these events can also be set to cause an interrupt and this is something we look at later in this chapter. For the moment, all we are concerned with is using these events as “memories” that something happened.

Currently MicroPython doesn't have any methods that support events, but it is fairly easy to add a pair of functions that do the job:

```
def gpio_get_events(pinNo):
    mask = 0xF << 4 * (pinNo % 8)
    intrAddr = 0x40014000 + 0x0f0 + (pinNo // 8)*4
    return (machine.mem32[intrAddr] & mask) >> (4 * (pinNo % 8))


def gpio_clear_events(pinNo, events):
    intrAddr = 0x40014000 + 0x0f0 + (pinNo // 8)*4
    machine.mem32[intrAddr] = events << (4 * (pinNo % 8))
```

How these work is explained in Chapter 18, for the moment all you really need to know is that `gpio_get_events` will return four bits that reflect the status of the events corresponding to `pinNo`. Similarly, `gpio_clear_events` is used to clear the latched events on `pinNo`. The latched events correspond to the constants:

```
Pin.IRQ_FALLING
Pin.IRQ_RISING
```

You can define more suitably named constants if you want to.

For example:

```
event = gpio_get_events(22)
if event & Pin.IRQ_RISING:
```

tests to see if a rising edge event has occurred on GP22.

Putting all this together, the steps in using events are:

1. Set the line to be an input:

   ```
   pin=Pin(22,Pin.IN,Pin.PULL_UP)
   ```

2. Clear the event:

   ```
   gpio_clear_events(22, Pin.IRQ_RISING)
   ```

3. After this you can do something else and eventually read the status bit to see if the event occurred in the intervening time, clearing the event to allow it to record another:

   ```
   event = gpio_get_events(22)
   gpio_clear_events(22, Pin.IRQ_RISING)
   ```

For example, to print which sort of latched event has occurred on GP22 you could use:

```
from utime import sleep
from machine import Pin
import machine

def gpio_get_events(pinNo):
    mask = 0xF << 4 * (pinNo % 8)
    intrAddr = 0x40014000 + 0x0f0 + (pinNo // 8)*4
    return (machine.mem32[intrAddr] & mask) >> (4 * (pinNo % 8))

def gpio_clear_events(pinNo, events):
    intrAddr = 0x40014000 + 0x0f0 + (pinNo // 8)*4
    machine.mem32[intrAddr] = events << (4 * (pinNo % 8))

pin=Pin(22,Pin.IN,Pin.PULL_UP)
while True:
    event=gpio_get_events(22)
    if(event & Pin.IRQ_FALLING):
        print("falling")
    if(event & Pin.IRQ_RISING):
        print("rising")
    gpio_clear_events(22, Pin.IRQ_FALLING | Pin.IRQ_RISING)
    sleep(0.5)
```

An Edgy Button

To make the difference between reading the line to detect a change of state and using the events, let's consider another version of our button program given earlier. In this case the GPIO line is set up for input and a message to press the button is printed. Then the program waits for 10 seconds and finally tests the state of the line. Even if the user has pressed the button lots of times during the pause, all that matters is the final state of the line as read when the `sleep(10)` ends:

```
from utime import sleep
from machine import Pin

pin=Pin(22,Pin.IN,Pin.PULL_DOWN)

print("Press Button")
sleep(10)
if pin.value():
    print("Button Pressed")
else:
    print("Button Not Pressed")
```

In other words, this program misses any button presses during the 10-second pause. This is a silly program, but now compare it to a version using edge events:

```
from utime import sleep
from machine import Pin
import machine

def gpio_get_events(pinNo):
    mask = 0xF << 4 * (pinNo % 8)
    intrAddr = 0x40014000 + 0x0f0 + (pinNo // 8)*4
    return (machine.mem32[intrAddr] & mask) >> (4 * (pinNo % 8))


def gpio_clear_events(pinNo, events):
    intrAddr = 0x40014000 + 0x0f0 + (pinNo // 8)*4
    machine.mem32[intrAddr] = events << (4 * (pinNo % 8))



pin=Pin(22,Pin.IN,Pin.PULL_DOWN)

print("Press Button")
gpio_clear_events(22, Pin.IRQ_FALLING)
sleep(10)
event = gpio_get_events(22)
gpio_clear_events(22, Pin.IRQ_FALLING)
if event & Pin.IRQ_FALLING:
    print("Button Pressed")
else:
    print("Button Not Pressed")
```

In this case the GPIO line is set up as an input with a pull-down and we test for a rising edge event rather than the line state. The difference is that if the user presses the button at any time during the 10-second sleep, the flag is set and the program registers the button press. The flag is set no matter what the program is doing, so instead of sleeping it could be getting on with some work, confident that it won't miss a button press. However, it cannot know exactly when the button was pressed and it cannot know how many times the button was pressed.

Events allow you to avoid missing a single input while polling, but cannot handle multiple inputs – if the user presses the button multiple times you still only detect a single press. You could implement a full queue-based event handling system, but this probably isn't worth the effort. A more reasonable alternative is to use the interrupt abilities of the GPIO lines, see later.

Measuring Pulses With Events

Now we have all of the functions we need to implement a pulse measurement program using events. In this case we can measure the width of any pulse as the distance between a rising and a falling edge or a falling and a rising edge. The main difference between this and the previous program that measured the width of a pulse is that now we are using the hardware to detect the state transitions, i.e. the "edges" of the signal.

To do this we need to detect the rising and falling edge for the pin.

The complete program is:

```
import time
from utime import sleep
from machine import Pin
import machine

def gpio_get_events(pinNo):
    mask = 0xF << 4 * (pinNo % 8)
    intrAddr = 0x40014000 + 0x0f0 + (pinNo // 8)*4
    return (machine.mem32[intrAddr] & mask) >> (4 * (pinNo % 8))


def gpio_clear_events(pinNo, events):
    intrAddr = 0x40014000 + 0x0f0 + (pinNo // 8)*4
    machine.mem32[intrAddr] = events << (4 * (pinNo % 8))

pin=Pin(22,Pin.IN,Pin.PULL_DOWN)
while True:
    gpio_clear_events(22, Pin.IRQ_FALLING | Pin.IRQ_RISING)
    while  not(gpio_get_events(22) & Pin.IRQ_RISING):
        pass
    t=time.ticks_us()
    while  not(gpio_get_events(22) & Pin.IRQ_FALLING):
        pass
    t=time.ticks_diff(time.ticks_us(),t)
    print(t)
    sleep(1)
```

After clearing both events we wait for a rising edge, clear it and then wait for a falling edge and take the time difference between the two.

This program produces very similar results to those of the previous program that simply read the inputs on the GPIO line. In this case there is no real advantage in using the events approach to polling as we are reading data as fast as we can anyway. However, if you had multiple GPIO lines and perhaps multiple conditions to test, you could set all the events you were interested in and then check to see if any of them had happened.

Interrupts

The Pico supports 32 distinct interrupts, but only 26 are actually used. All of the user GPIO lines act together to create a single IO interrupt. A subtle point is that each of the Pico's two processors can respond at the same time to an IO interrupt caused by a different GPIO line – that is, the IO interrupts are not shared between cores. In all other cases interrupts can only be enabled on one core at a time.

The fact that there is only a single interrupt for all of the GPIO lines means it is up to the interrupt handler to work out which line caused the interrupt and to reset it after dealing with it. Which GPIO line can generate an interrupt is specified by the same bits in the same register that we have been using as event indicators in the earlier sections. If an interrupt is enabled for a given line and a given event then the interrupt will occur if that bit is set to one.

The events that you can use are the same as before:

```
GPIO_IRQ_LEVEL_LOW
GPIO_IRQ_LEVEL_HIGH
GPIO_IRQ_EDGE_FALL
GPIO_IRQ_EDGE_RISE
```

MicroPython provides a simple method for working with the edge GPIO interrupts, level interrupts are not implemented:

```
pin.irq(ISR, edge)
```

This sets `ISR` as the interrupt service handler for the edge event. This is a very simplified version of what happens at a lower level. The `ISR` function has to accept a single parameter, which is a Pin object associated with the GPIO line that caused the event.

There are, in fact, two distinct implementations of an interrupt system within MicroPython. The first is the default and it seems to be a full interrupt queue implementation, which is good if you don't want to miss an interrupt, but not so good for timing. The second is a lighter wrapping of the interrupt mechanism which doesn't allow an interrupt to occur within the interrupt handler and is more accurate from the point of view of timing. To use the lighter interrupt handling you have to add `hard=True` to the `irq` method:

```
pin.irq(ISR, edge,hard=True)
```

Unfortunately, neither form of interrupt returns a timestamp for when the interrupt occurs, which limits their usefulness. The difference between the two types of interrupt handling is explored more in the next section.

Notice that the `ISR` will still be called even if your MicroPython program has come to an end:

```
def HelloIRQ(pin):
    print("IRQ")

pin=Pin(22,Pin.IN,Pin.PULL_DOWN)
pin.irq(HelloIRQ,Pin.IRQ_RISING)
```

You will see `IRQ` printed whenever GP22 has a rising edge, even after the program has finished.

Although the hardware only supports a single ISR for all of the GPIO lines, MicroPython implements a system that lets you associate a different ISR for each GPIO line. For example:

```
def HelloIRQ1(pin):
    print("IRQ1")

def HelloIRQ2(pin):
    print("IRQ2")

pin1=Pin(21,Pin.IN,Pin.PULL_DOWN)
pin2=Pin(22,Pin.IN,Pin.PULL_DOWN)

pin1.irq(HelloIRQ1,Pin.IRQ_RISING)
pin2.irq(HelloIRQ2,Pin.IRQ_RISING)
```

You will see `IRQ1` or `IRQ2` printed depending on which pin the interrupt occurred on. Also notice that the implementation automatically clears the interrupt and disables interrupts while in the ISR. This makes using MicroPython interrupts much easier, but it also slows down the basic IRQ handling from what is possible using the raw hardware.

There are also some other restrictions on writing an interrupt service routine (ISR). It should be as fast as possible and this implies that it shouldn't allocate memory, i.e. create or extend objects. Preferably create all objects that the ISR needs to use before it runs and, if you need to add to an object, find a representation that allows you to allocate the space needed before using it. For example, prefer an `Array.array` object to a `List` or `Dictionary` object as an `Array.array` object is preallocated at its initialized size. These restrictions are advisory for the default interrupt implementation, but mandatory for the hard version.

This is all there is to using GPIO to generate an interrupt and handling it. In practice, things are more difficult to get right than you might expect. There is the question of what happens if an interrupt occurs while you are in the interrupt handler and what about accessing shared resources while in the interrupt handler?

How Fast Is An Interrupt?

Let's find out how much overhead is inherent in an interrupt by repeating the pulse width measurement. This time, we can't simply print the results as this would stop the interrupt-handling. As a compromise, we save 20 readings in an array and then print them. It is also important to keep the interrupt-handling routines short as how long they take to complete governs how fast the next interrupt can be handled. The difference between the two interrupt implementations is great and they both have serious limitations.

The full program for the first implementation is:

```
import time
from machine import Pin
import machine
import array

count = 0
t = array.array('L', [0]*20)

def myHandler(pin):
    global t, count
    t[count] = time.ticks_us()
    count = count+1
    if count > 19:
        for i in range(1, 20):
            print(time.ticks_diff(t[i], t[i-1]))
        pin.irq(None, Pin.IRQ_RISING, hard=True)
        return

pin = Pin(22, Pin.IN, Pin.PULL_DOWN)
pin.irq(myHandler, Pin.IRQ_RISING, hard=False)
```

Notice that the `if` statement not only prints the data, it also disables the interrupts by setting the IRQ handler to `None`. If you run the program with an input frequency that should give 1000μs, you will see results something like:

```
390
79
915
. . .
930
1000
1003
998
1001
Traceback (most recent call last):
  File "<stdin>", line 16, in myHandler
...repeated some number of times
```

There are two things to notice. The first is that many of the measurements are very low. The reason for this is likely to be the time it takes to respond to the interrupt. That is, the time between the interrupt occurring and the interrupt routine being called is on average very long. The second is that the error message occurs because interrupts are still being serviced while the interrupt handler is running. What seems to be happening is that each interrupt that occurs while the interrupt handler is running is added to a queue and then processed after the interrupts have been turned off – hence the array overrun. This makes it less likely you will miss an interrupt, but as there is no interrupt timestamp it makes it virtually impossible to do this sort of timing. That is, interrupts are not missed, but when the interrupt handler is called relative to when the interrupt occurred is vague.

If you change the interrupt to `hard=True`:

```
pin.irq(myHandler, Pin.IRQ_RISING, hard=True)
```

then the results are typically very different:

```
972
970
995
1003
998
989
1010
1003
. . .
1003
998
1001
998
```

Now the times are closer to the 1000μs you would expect and there are no error messages. This means that the interrupt handler is being called closer to the time that the interrupt actually happened and interrupt handling is switched off during the interrupt handler. This means you might well miss an interrupt, but the timings are more accurate. However, the time between the interrupt occurring and the interrupt handler being called is still about 20μs.

In conclusion, neither type of interrupt handling is suitable for timing events. The addition of a timestamp would improve this.

Race Conditions and Starvation

One of the big problems of using interrupts on a small processor like the Pico is that there is no operating system to make sure that different tasks get a fair share of the processor. It is entirely possible for an interrupt routine to be called so often that the main program, or some other function, doesn't get a chance to execute. For example, if you try a program that prints a message in the main program and responds to interrupts then you can see the starvation of the main program's code very easily:

```
import time
from machine import Pin

def myHandler(pin):
    print(time.ticks_us())

pin = Pin(22, Pin.IN, Pin.PULL_DOWN)
pin.irq(myHandler, Pin.IRQ_RISING, hard=True)
while True:
    print("doing something useful")
```

If you apply a square wave input to GP22 then you will see both the "doing something" and the interrupt message, as long as the frequency is 200Hz or less. Above this frequency the interrupt routine is called almost at once after it returns to the main program. As a result the main program makes little or no progress – it is starved of CPU time. The reason for the very low interrupt rate that causes starvation is due to the time the interrupt routine takes to run as printing is slow. A faster routine would work at a higher frequency before starvation set in, but it is a constant problem with interrupt-driven systems. If the interrupt occurs fast enough you will even lose control of the Pico because there is not time for it to communicate with the outside world – the solution is to power cycle it.

If you also look closely at the output of the program when the main program is getting a chance to run, you will see the second problem with simple interrupt systems – race conditions. The output isn't clean, instead it is all jumbled together:

```
doing something useful
doing30948911
 something useful
doing something useful
doing something useful
```

The reason for this is simply that the interrupt can occur while the program is in the middle of a call to `print`. That is, the call to `print` is not "atomic" and can be split by the action of the interrupt. The key takeaway is that interrupts can occur anywhere within your program, even within actions that you think of as atomic, i.e. which cannot be interrupted.

It is very easy to create an interrupt-driven program that locks up the Pico and needs a power cycle to restart. The only reasonably safe use of Pico interrupts is when the events are very slow and there is plenty of time to process them.

You can use:

```
s=machine.disable_irq()
print("doing something useful")
machine.enable_irq(s)
```

to create sections of code where interrupts cannot occur, but notice that this disables all interrupts, including timer ticks.

Responding To Input

This look at methods of dealing with the problems of input isn't exhaustive – there are always new ways of doing things, but it does cover the most general ways of implementing input. As already mentioned, the problem with input is that you don't know when it is going to happen. What generally matters is speed of response.

For low-frequency inputs, interrupts are worthwhile. They can leave your program free to get on with other tasks and simplify its overall structure. For high-frequency inputs that need to be serviced regularly, a polling loop is still the best option for maximum throughput. How quickly you can respond to an input depends on how long the polling loop is and how many times you test for it per loop.

Summary

- Events are a stored indication that something happened.

- Interrupts are events that cause something to happen.

- You can use an event with a polling loop to protect against missing input because the program is busy doing something else.

- If an event occurs before the current event has been cleared then it might be missed. To avoid missing events, you can use an event queue which stores events in the order they happened until they are processed.

- MicroPython doesn't support events, but it is easy to add a function to make use of them. You can enable a bit to be set if a GPIO line goes high, low or is high or low. Edge-triggered events are the easiest to work with.

- Using events in a polling loop hardly slows things down at all.

- You can use events to generate an interrupt and call an interrupt handler.

- MicroPython has two interrupt implementations. The default, `hard=False` allows interrupts to occur during the interrupt handler and a queue of interrupts is formed. The alternative, `hard=True` disables interrupts while in the interrupt handler and doesn't keep a queue.

- Both interrupt implementations have a latency of at least 20μs and usually more. Only `hard=True` interrupt handling has any chance of responding to an interrupt close to the time it occurred.

- Whenever possible, avoid using interrupts.

Chapter 8
Pulse Width Modulation

One way around the problem of getting a fast response from a microcontroller is to move the problem away from the processor. In the case of the Pico there are some built-in devices that can use GPIO lines to implement protocols without the CPU being involved. In this chapter we take a close look at the use of Pulse Width Modulation (PWM) including generating sound, driving LEDs and servos.

When performing their most basic function, i.e. output, the GPIO lines can be set high or low by the processor. How fast they can be set high or low depends on the speed of the processor.

Using the GPIO line in its Pulse Width Modulation (PWM) mode you can generate pulse trains up to 60MHz. The reason for the increase in speed is that the GPIO is connected to a pulse generator and, once set to generate pulses of a specific type, the pulse generator just gets on with it, without needing any intervention from the GPIO line or the processor. In fact, the pulse output will continue after your program has ended. Of course, even though the PWM line can generate pulses very fast pulses, usually what you want to do is change the nature of the pulses and this is a slower process involving the processor.

Some Basic Pico PWM Facts

There are some facts worth getting clear right from the start, although some of their significance will only become clear as we progress.

First, what is PWM? The simple answer is that a pulse width modulated signal has pulses that repeat at a fixed rate, say one pulse every millisecond, but the width of the pulse can be changed.

There are two basic things to specify about the pulse train that is generated, its repetition rate and the width of each pulse. Usually the repetition rate is set as a simple repeat period and the width of each pulse is specified as a percentage of the repeat period, referred to as the duty cycle.

So, for example, a 1ms repeat and a 50% duty cycle specifies a 1ms period, which is high for 50% of the time, i.e. a pulse width of 0.5ms. The two extremes are 100% duty cycle, i.e. the line is always high, and 0% duty cycle, i.e. the line is always low.

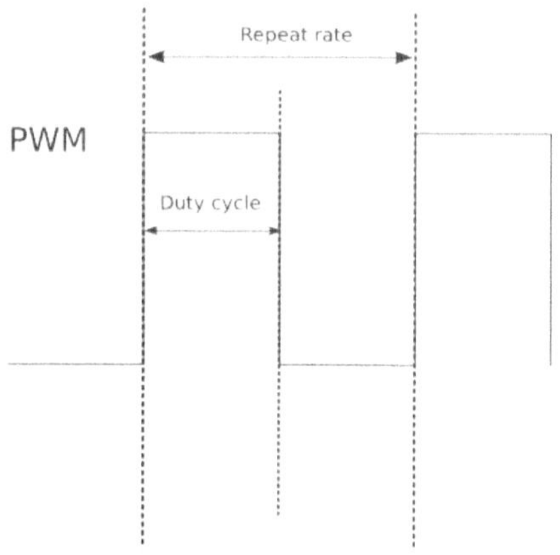

Notice it is the duty cycle that carries the information in PWM and not the frequency. What this means is that, in general, you select a repeat rate and stick to it and what you change as the program runs is the duty cycle.

In many cases PWM is implemented using special PWM-generator hardware that is either built into the processor chip or provided by an external chip. The processor simply sets the repeat rate by writing to a register and then changing the duty cycle by writing to another register. This ideally provides the best sort of PWM with no load on the processor and glitch-free operation. You can even buy add-on boards that will provide additional channels of PWM without adding to the load on the processor.

The alternative to dedicated PWM hardware is to implement it in software. You can work out how to do this quite easily. All you need is a timing loop to set the line high at the repetition rate and then set it low again according to the duty cycle. You can implement this using either interrupts or a polling loop and in more advanced ways, such as using a DMA (Direct Memory Access) channel.

Pico PWM

In the case of the Pico, the PWM lines are implemented using special PWM hardware. It has eight PWM generators, each capable of two PWM outputs. Any of the GPIO lines can be used as PWM lines and this means you can have up to 16 PWM lines in operation at any given time. Things are a little more complicated in that each pair of outputs has the same frequency, which means you have eight, independently set, pairs of outputs. In addition, one of the outputs can be used as an input and this reduces the number of outputs available.

The PWM generators are assigned to GPIO pins in a fixed order:

GPIO	0	1	2	3	4	5	6	7	8	9	10	11	12	13	14	15
PWM Channel	0A	0B	1A	1B	2A	2B	3A	3B	4A	4B	5A	5B	6A	6B	7A	7B

GPIO	16	17	18	19	20	21	22	23	24	25	26	27	28	29		
PWM Channel	0A	0B	1A	1B	2A	2B	3A	3B	4A	4B	5A	5B	6A	6B		

You don't have to know about how the PWM hardware works, but it helps with understanding some of the restrictions.

To create a `PWM Pin` object you have to pass its constructor a `Pin` object. For example:

```
pwm16 = PWM(Pin(16))
```

creates a `PWM` object associated with GP16. You can set the frequency using:

```
pwm16.freq(500)
```

which sets the frequency in Hz. The PWM hardware isn't enabled at this point. To start it generating a signal you have to set the duty cycle. This is done using:

```
pwm16.duty_u16((duty)
```

where `duty` is a value in the range 0 to 65,535 corresponding to 0 to 100%.

There is also:

```
pwm16.duty_ns(ns)
```

which sets the time the line is high in nanoseconds. This isn't as useful for general use and if you specify a time that is greater than the set period you will generate an exception.

You can easily create your own duty cycle methods that work in terms of percentages or whatever way you want to specify the duty cycle. For example, you can create a new class which has a duty cycle set as a percentage:

```
class myPWM(PWM):
    def __init__(self, pin: Pin):
        super().__init__(pin)
    def duty(self,d):
        print(65535*d//1000)
        super().duty_u16(65535*d//1000)
```

In this case the percentage is specified multiplied by 10. For example, to set a 50% duty cycle you would use:

```
pwm16 = myPWM(Pin(16))
pwm16.freq(250)
pwm16.duty(500)
```

Once you have set the duty cycle the PWM generator starts to output the specified PWM signal on the pin. Notice that this works even after your program has completed. To stop the PWM signal use:

```
pwm16.deinit()
```

You can change the frequency or duty cycle at any time.

Although you don't need to know anything about the PWM hardware, there is one limitation you need to be aware of. Each PWM generator supports two outputs which work at the same frequency and optionally different duty cycles. For example, if you look back to the table of PWM pin assignments you can see that PWM 0 has outputs on GP16 and GP17. This means that these two pins can only work at the same frequency. For example:

```
from machine import Pin, PWM
pwm16 = PWM(Pin(16))
pwm17 = PWM(Pin(17))

pwm16.freq(250)

pwm16.duty_u16(65535//2)
pwm17.duty_u16(65535//4)
```

produces a PWM signal on GP16 and GP17 with the same frequency of 250Hz and duty cycles of 50% and 25% respectively. If you change the frequency on either pin, both change frequency.

You can see the result in this logic analyzer display:

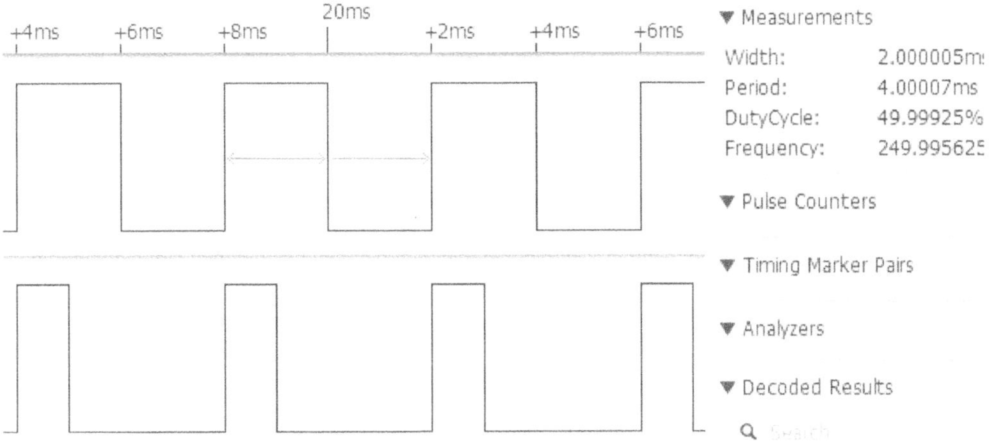

You can see from the logic analyzer trace that the pulses on each line start their duty cycle at exactly the same time. The PWM hardware can create phase-correct pulses where the pulses are aligned about their center point, but MicroPython doesn't support this mode. However, it is fairly easy to write a function that sets or unsets phase-correct mode:

```
def pwm_set_phase(sliceNo,phase):
    Addr = 0x40050000 +0x14*sliceNo
    if phase:
        machine.mem32[Addr]=machine.mem32[Addr] | 0x2
    else:
        machine.mem32[Addr]=machine.mem32[Addr] & 0xFFFFFFFD
```

Using this we can generate phase-correct pulses:

```
pwm16 = PWM(Pin(16))
pwm17 = PWM(Pin(17))
pwm16.freq(250)
pwm_set_phase(0,True)
pwm16.duty_u16(65535//2)
pwm17.duty_u16(65535//4)
```

Now you can see that the pulses don't start at the same time, but they are centered around the same time:

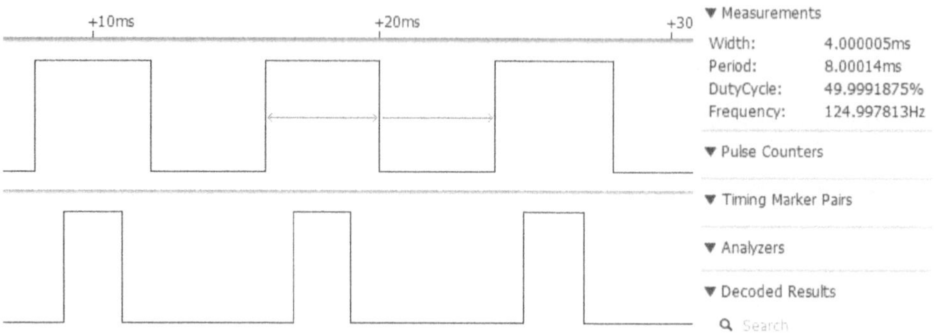

As well as being able to set the level for each channel, you can also set the polarity. However, it is easy to write a function that will invert any channel:

```
def pwm_set_polarity(sliceNo,channel,invert):
    Addr = 0x40050000 +0x14*sliceNo
    if invert:
       machine.mem32[Addr]=machine.mem32[Addr] | 0x1 << (2+channel)
    else:
       machine.mem32[Addr]=machine.mem32[Addr] & ~(0x1<<(2+channel))
```

For example:

```
pwm16 = PWM(Pin(16))
pwm17 = PWM(Pin(17))

pwm16.freq(250)
pwm_set_polarity(0,1,True)
pwm16.duty_u16(65535//4)
pwm17.duty_u16(65535//4)
```

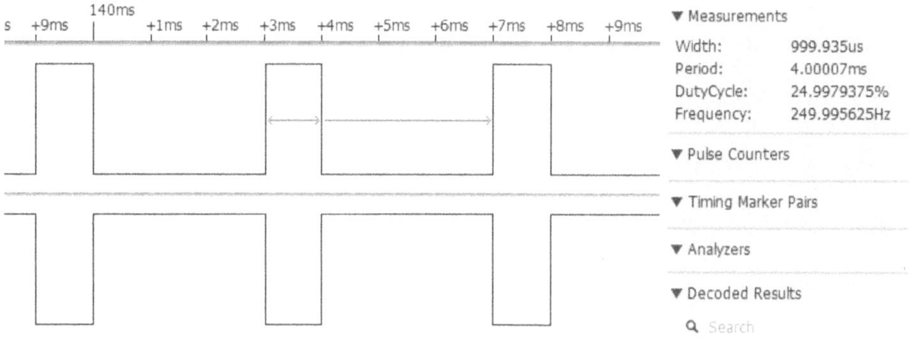

and you can see that the output of channel B is inverted.

Changing The Duty Cycle

For reasons that will be discussed later, in most cases the whole point is to vary the duty cycle or the period of the pulse train. This means that the next question is, how fast can you change the characteristics of a PWM line? In other words, how fast can you change the duty cycle? There is no easy way to give an exact answer and, in most applications, an exact answer isn't of much value. The reason is that for a PWM signal to convey information it generally has to deliver a number of complete cycles with a given duty cycle. This is because of the way pulses are often averaged in applications.

We also have another problem – synchronization. This is more subtle than it first seems. The hardware won't change the duty cycle until the current pulse is complete. You might think that the following program works to switch between two duty cycles on a per pulse basis:

```
pwm16 = PWM(Pin(16))

pwm16.freq(50)

pwm16.duty_u16(65535//2)
while True:
    pwm16.duty_u16(65535//2)
    pwm16.duty_u16(65535//4)
```

but if you try this out the result isn't what you might expect on a first analysis:

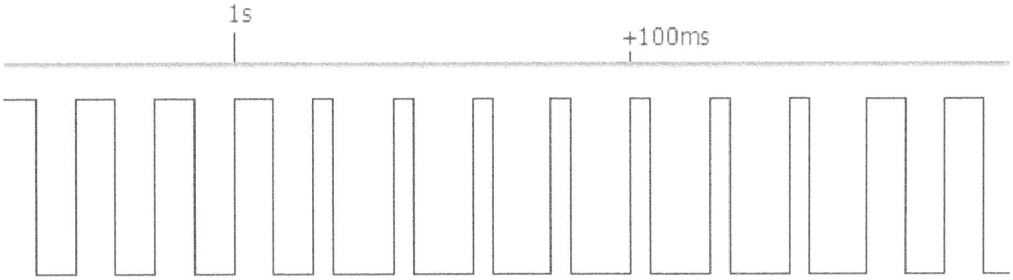

You don't get one 25% followed by one 50% pulse, but a varying number of each in turn. The reason is, of course, that the duty cycle is being set asynchronously.

Duty Cycle Resolution

It seems as if you can set the duty cycle to a 16-bit value giving you 65535 different settings, however, this is only true in a few cases. The Pico's PWM hardware implements the following algorithm:

```
set GPIO line high
for count in range(wrap):
       wait one clock cycle
       if count>level:
              set GPIO line low
```

In other words, the frequency of the PWM signal depends on the value of `wrap` and the clock frequency and duty cycle depend on `level`. When you set the frequency of the PWM signal that you want, the software works out the best value for the clock frequency and wrap value. In an ideal wold the wrap would always be the maximum 16-bit value, 65535, but to achieve the desired frequency it often has to be less. Clearly you can only set the duty cycle between `0` and `wrap` and this can limit the precision that you can set.

For example, if `wrap` is 4 then you can only set duty cycles of 0, 25%, 75% and 100%, i.e. `level= 0, 1, 2, 3`, and such low values of `wrap` do occur.

In other words, the frequency you select affects the accuracy of the setting of the duty cycle. In most cases the wrap value is large enough to ignore the problem, but sometimes it is important. MicroPython doesn't provide any way to explicitly set or get the wrap value, but it is easy to add a function to get it:

```
def pwm_get_wrap(sliceNo):
    Addr = 0x40050000 +0x10+0x14*sliceNo
    return (machine.mem32[Addr])
```

Problems with small wrap values usually occur at high frequencies. The reason is that the clock runs at 125MHz and so if you want a PWM frequency of say 30MHz then the wrap value has to be 4 and hence the duty cycle can only be `0,1,2,3`. You can see that this is the case in the following example:

```
pwm16 = PWM(Pin(16))
pwm16.freq(125000000//4)
print(pwm_get_wrap(0))
pwm16.duty_u16(65535//2)
```

You will see the wrap is 2 and if you examine the output using a logic analyzer you will see that the duty cycle is not 50%.

At lower frequencies the duty cycle precision is usually close to 16-bits. Also notice that the duty cycle is always set as if it was a full 16-bit value and MicroPython scales this into the range that wrap makes possible.

There is an additional effect related to the clock that you need to consider. The clock divider can be fractional and in this case the fractional division is achieved by varying the division between two integer dividers so that the average frequency is as requested. This can lead to jitter in the pulse positions.

Uses Of PWM – Digital To Analog

What sorts of things do you use PWM for? There are lots of very clever uses for PWM. However, there are two use cases which account for most PWM applications - voltage or power modulation and signaling to servos.

The amount of power delivered to a device by a pulse train is proportional to the duty cycle. A pulse train that has a 50% duty cycle is delivering current to the load only 50% of the time and this is irrespective of the pulse repetition rate. So the duty cycle controls the power, but the period still matters in many situations because you want to avoid any flashing or other effects. A higher frequency smooths out the power flow at any duty cycle.

If you add a low-pass filter to the output of a PWM signal then what you get is a voltage that is proportional to the duty cycle. This can be looked at in many different ways, but again it is the result of the amount of power delivered by a PWM signal. You can also think of it as using the filter to remove the high-frequency components of the signal, leaving only the slower components due to the modulation of the duty cycle.

How fast you can work depends on the duty cycle resolution. If you work with 8-bit resolution your D-to-A conversion will have 256 steps, which at 3.3V gives a potential resolution of 3.3/256 or about 13mV. This is often good enough. If you understand the previous section on duty cycle resolution this means that the PWM frequency has to 125000000//256 to give a wrap of 256. The PWM output in this configuration mimics the workings of an 8-bit A-to-D converter.

To demonstrate the sort of approach that you can take to D-to-A conversion, the following program creates a sine wave. To do this we need to compute the duty cycle for 256 points in a complete cycle. We could do this each time a value is needed, but to make the program fast enough we have to compute the entire 256 points and store them in an array. While there are fixed-point arithmetic ways of computing the sine or the cos of an angle, it is simpler to use the Pico's built-in floating-point software. Notice that the Pico doesn't have floating-point arithmetic implemented in hardware and it is best to avoid it if possible.

The program is:

```
from machine import Pin, PWM
import array
import machine
import math


def pwm_get_wrap(sliceNo):
    Addr = 0x40050000 + 0x10+0x14*sliceNo
    return (machine.mem32[Addr])


wave = array.array('H', [0]*256)
for i in range(256):
    wave[i] = int(65535//2 + (math.sin(i * 2.0 * 3.14159 / 256.0) *
65535//2))


pwm16 = PWM(Pin(16))
pwm16.freq(125000000//256)

print(pwm_get_wrap(0))
while(True):
    for i in range(256):
        pwm16.duty_u16(wave[i])
```

The 16-bit duty cycle values needed are computed and stored in the `wave` array. Then the `PWM` is set up with the highest frequency that will provide 256 levels of duty cycle. Finally, a `for` loop is used to set the duty cycle from the array. Notice that there is no attempt at synchronizing when the duty cycle update will occur – it happens as fast as possible.

The waveform repeats after around 8.8ms, which makes the frequency around 113Hz.

To see the analog waveform, we need to put the digital output into a low-pass filter. A simple resistor and capacitor work reasonably well:

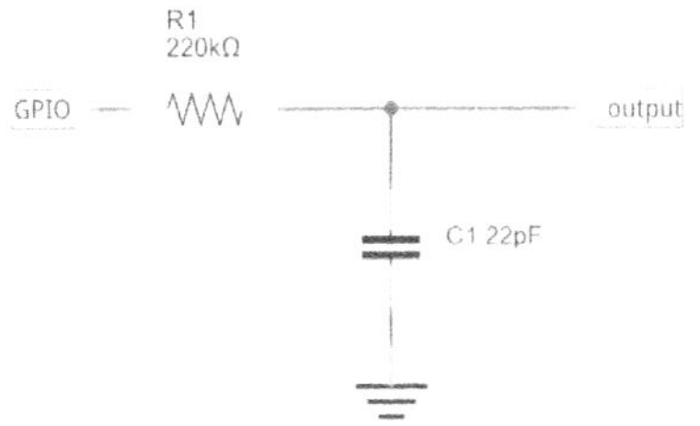

The filter's cutoff is around 33kHz and might be a little on the high side for this low-frequency, 200Hz, output, but it produces a reasonable waveform:

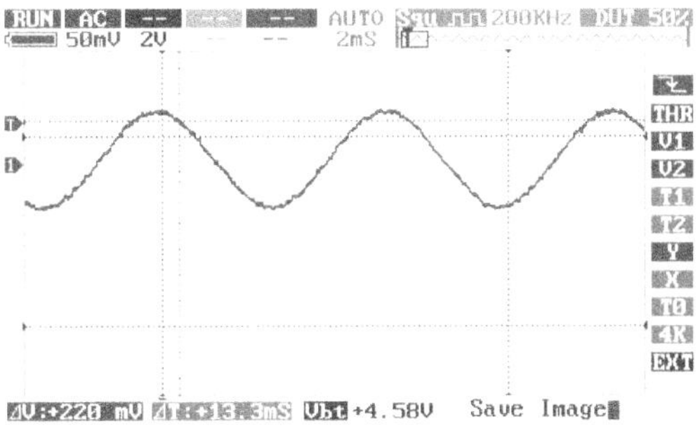

You can use this technique to create a sine wave, or any other waveform you need, but for high-quality audio you need a higher sampling rate.

Frequency Modulation

Using PWM to create musical tones, and sound effects in general, is a well-explored area which is too wide to cover in this book. In most cases we choose to vary the duty cycle at a fixed sample rate, but an alternative is to leave the duty cycle fixed at, say, 50% and modulate the frequency. You can use this approach to create simple musical tones and scales.

As the frequency of middle C is 281.6Hz, to generate middle C you could use:

```
pwm16 = PWM(Pin(16))
pwm16.freq(282)
pwm16.duty_u16(65535//2)
```

The resulting output is a square wave with a measured frequency of 282.1Hz which isn't particularly nice to listen to. You can improve it by feeding it through a simple low-pass filter like the one used above for waveform synthesis. You can look up the frequencies for other notes and use a table to generate them.

Controlling An LED

You can also use PWM to generate physical quantities such as the brightness of an LED or the rotation rate of a DC motor. The only differences required by these applications are to do with the voltage and current you need and the way the duty cycle relates to whatever the physical effect is. In other words, if you want to change some effect by 50%, how much do you need to change the duty cycle? For example, how do we "dim" an LED?

The simplest example is to drive the on-board LED using a PWM signal:

```
from machine import Pin, PWM
from machine import Pin, PWM
pwm25 = PWM(Pin(25))
pwm25.freq(2000)

while True:
    for d in range(0,65535,655):
        pwm25.duty_u16(d)
        sleep_ms(50)
```

If you try this out you will see the LED slowly increase in brightness, but it seems to be a longer time at maximum brightness than at any other value. This is a consequence of the non-linear relationship between duty cycle and perceived brightness.

By changing the duty cycle of the PWM pulse train you can set the amount of power delivered to an LED, or any other device, and hence change its brightness. If you use a 50% duty cycle, the LED is on 50% of the time and it has been determined that this makes it look as if it is half as bright. However, this is not the end of the story as humans don't respond to physical brightness in a linear way. The Weber-Fechner law gives the general relationship between perceived intensity and physical stimulus as logarithmic. In the case of an LED, the connection between duty cycle and brightness is a complicated matter, but the simplest approach uses the fact that the perceived brightness is roughly proportional to the cube root of the physical brightness. The exact equations, published as CIE 1931, are:

$L = 903.3 \cdot (Y / Y_n) \qquad (Y/ Y_n) \leq 0.008856$

$L = 116 \cdot (Y / Y_n)^{1/3} - 16 \qquad (Y/ Y_n) > 0.008856$

where L is the perceived brightness and Y / Y_n is a measure of physical brightness.

The exact relationship is complicated, but in most cases a roughly cubic law, obtained by inverting the CIE relationship, can be used:

$d = kb^3$

where b is the perceived brightness and d is the duty cycle. The constant k depends on the LED. The graph below shows the general characteristic of the relationship for a duty cycle of 0 to 100% on the y-axis and arbitrary, 0 to 100, perceived brightness units on the x-axis.

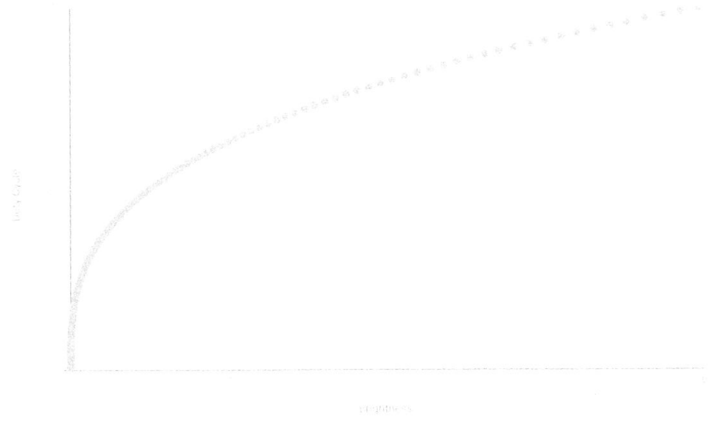

Notice that, as the LED when powered by a PWM signal is either full on or full off, there is no effect in the change in LED light output with current - the LED is always run at the same current. What all of this means is that if you want an LED to fade in a linear fashion you need to change the duty cycle in a non-linear fashion. Intuitively it means that changes when the duty cycle is small produce bigger changes in brightness than when the duty cycle is large.

A program to implement cubic dimming is:

```
from utime import sleep_ms
from machine import Pin, PWM

pwm25 = PWM(Pin(25))
pwm25.freq(2000)

while True:
    for b in range(0,100):
        pwm25.duty_u16(int(65535*b*b*b/1000000))
        sleep_ms(50)
```

If you try this out you should notice that the LED changes brightness more evenly across its range. The only problem is that now 100 steps are insufficient to mask the steps in brightness at the lower level. The solution is to work with a more precise specification of duty cycle.

In most cases it is irrelevant exactly how linear the response of the LED is - a rough approximation looks as smooth to the human eye. You can even get away with using a square law to dim the LED. The only exception is when you are trying to drive LEDs to create a gray-level or color display when color calibration is another level of accuracy.

There is also the question of what frequency we should use. Clearly it has to be fast enough not to be seen as flickering and this generally means it has to be greater than 80Hz, the upper limit for human flicker fusion, but, because of the strobe effect, flickering becomes more visible with moving objects. The faster the LED switches on and off, the less flicker should be visible, but before you select frequencies in the high kHz range it is worth knowing that an LED has a minimum time to turn on and so frequencies at low kHz work best.

If you want to dim something other than the on-board LED, you will, in many cases, need a driver to increase the brightness. For a simple example, consider connecting a standard LED to the PWM line and using the BJT driver circuit introduced in Chapter 5.

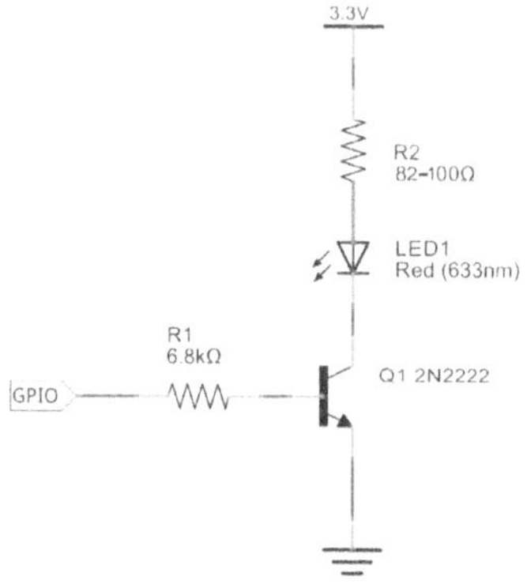

What Else Can You Use PWM For?

PWM lines are incredibly versatile and it is always worth asking the question "could I use PWM?" when you are considering almost any problem. The Pico's PWM generator is particularly versatile. For example, it has an input option which can be used to determine the duty cycle of an input wave form. However, MicroPython doesn't support this and many of its more advanced features are hidden from you for the sake of simplicity. For most of the time this tradeoff is worth it.

The LED example suggests how you can use PWM as a power controller. You can extend this idea to a computer-controlled switch-mode power supply. All you need is a capacitor to smooth out the voltage and perhaps a transformer to change the voltage. You can also use PWM to control the speed of a DC motor and, by adding a simple bridge circuit, you can control its direction and speed. Finally, you can use a PWM signal as a modulated carrier for data communications. For example, most infrared controllers make use of a 38kHz carrier, which is roughly a 26μs pulse. This is switched on and off for 1ms and this is well within the range that the PWM can manage. So all you have to do is replace the red LED in the previous circuit with an infrared LED and you have the start of a remote control, or data transmission, link.

One big area of use is in controlling motors, and servo motors in particular, and this is the subject of the next chapter.

Summary

- PWM, Pulse Width Modulation, has a fixed repetition rate but a variable duty cycle, i.e. the amount of time the signal is high or low changes.
- PWM can be generated by software simply by changing the state of a GPIO line correctly, but it can also be generated in hardware so relieving the processor of some work.
- As well as being a way of signaling, PWM can also be used to vary the amount of power or voltage transferred. The higher the duty cycle, the more power/voltage.
- The Pico has eight hardware PWM generators and these are capable of a range of operational modes.
- The PWM lines are controlled by a counter and two values wrap which gives the frequency and level which gives the duty cycle. This is mostly hidden from the MicroPython programmer and you can specify frequency and duty cycle without worrying too much about how they are generated.
- You can generate phase-correct PWM or allow the phase to vary with the duty cycle.
- The higher the wrap value, the higher the resolution of the duty cycle. It is possible to work out the best value for the clock frequency for any PWM frequency to maximize the duty cycle resolution.
- PWM can be used to implement digital to analog conversion simply by varying the duty cycle. In the same way, by varying the duty cycle, you can dim an LED.

Chapter 9

Controlling Motors And Servos

Controlling motors is an obvious use for the low cost Pico, but it is important to understand the different types of motor that you can use and exactly how to control them using PWM. In addition to PWM control, we also look at the very useful stepper motor which doesn't make use of PWM.

The simplest division among types of motor is AC and DC. AC motors are generally large and powerful and run from mains voltage. As they are more difficult to work with, and they work at mains voltages, these aren't used much in IoT applications. DC motors generally work on lower voltage and are much more suitable for the IoT. In this chapter we will only look at DC motors and how they work thanks to pulse width modulation. The parts used are listed in the Resources section of the book's webpage at www.iopress.info.

DC Motor

There are two big classes of DC motor – brushed and brushless. All motors work by using a set of fixed magnets, the stator, and a set of rotating magnets, the rotor. The important idea is that a motor generates a "push" that rotates the shaft by the forces between the magnet that makes up the stator and the magnet that makes up the rotor. The stronger these magnets are, the stronger the push and the more torque (turning force) the motor can produce. To keep the motor turning, one of the two magnetic fields has to change to keep the rotor being attracted to a new position.

DC motors differ in how they create the magnetism in each component, either using a permanent magnet or an electromagnet.

This means there are four possible arrangements:

	1	2	3	4
Stator	Permanent	Permanent	Electromagnet	Electromagnet
Rotor	Permanent	Electromagnet	Permanent	Electromagnet
Type	Can't work	Brushed DC	Brushless DC	Series or shunt

Arrangement 1 can't produce a motor because there is no easy way of changing the magnetic field. Arrangement 4 produces the biggest and most powerful DC motors used in trains, cars and so on. Arrangement 2, Brushed DC, is the most commonly encountered form of "small" DC motor. However, arrangement 3, brushless DC, is becoming increasingly popular.

Different arrangements produce motors which have different torque characteristics, i.e. how hard they are to stop at any given speed. Some types of motor are typically low torque at any speed, i.e. they spin fast but are easy to stop.

Low torque motors are often used with gearboxes, which reduce the speed and increase the torque. The big problem with gearboxes, apart from extra cost, is backlash. The gears don't mesh perfectly and this looseness means that you can turn the input shaft and at first the output shaft won't move. Only when the slack in the gears has been taken up will the output shaft move. This makes a geared motor less useful for precise positioning, although there are ways to improve on this using feedback and clever programming.

Brushed Motors

To energize the electromagnets, a brushed motor supplies current to the armature via a split ring or commutator and brushes. As the rotor rotates, the current in the coil is reversed and it is always attracted to the other pole of the magnet.

The only problem with this arrangement is that, as the brushes rub on the slip ring as the armature rotates, they wear out and cause sparks and hence RF interference. The quality of a brushed motor depends very much on the design of the brushes and the commutator.

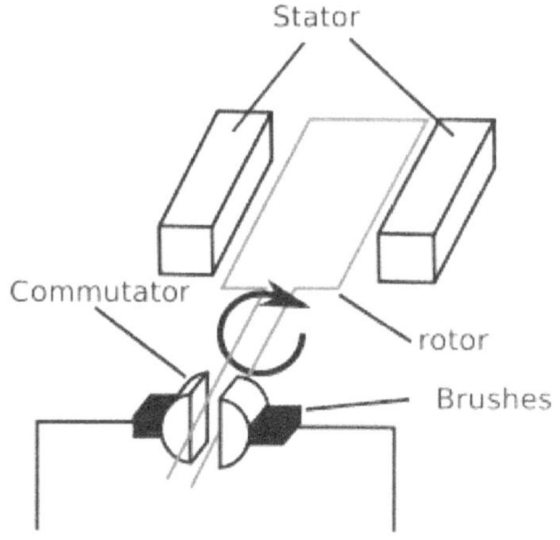

Very small, cheap, brushed DC motors, of the sort in the picture below, tend to not have brushes that can be changed and when they wear out the motor has to be replaced. They also tend to have very low torque and high speed. This usually means that they have to be used with a gearbox. If you overload a brushed motor then the tendency is to demagnetize the stator magnets. The cheapest devices are basically toys.

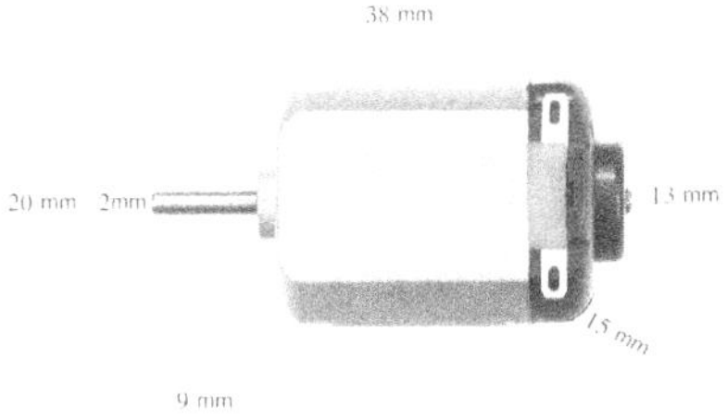

Higher quality brushed motors are available and they also come in a variety of form factors. For example, the 775 motor is 66.7 by 42mm with a 5mm shaft:

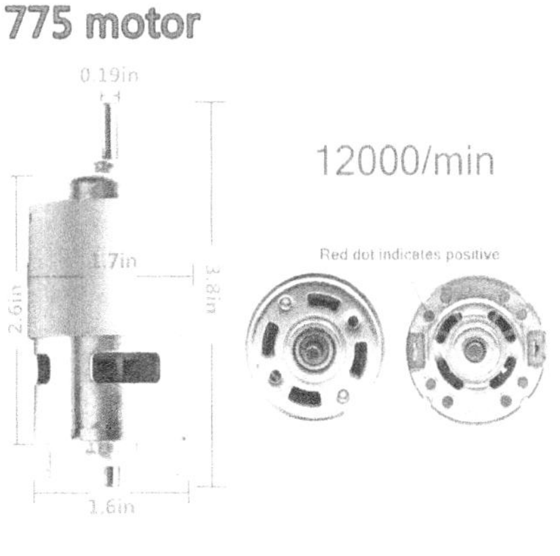

Even these motors tend not to have user-serviceable brushes, but they tend to last a long time due to better construction.

Unidirectional Brushed Motor

A brushed motor can be powered by simply connecting it to a DC supply. Reversing the DC supply reverses the direction of the motor. The speed is simply proportional to the applied voltage. If all you want is a unidirectional control then all you need is a PWM driver that can supply the necessary current and voltage.

A single transistor solution is workable as long as you include a diode to allow the energy stored in the windings to discharge when the motor is rotating, but not under power:

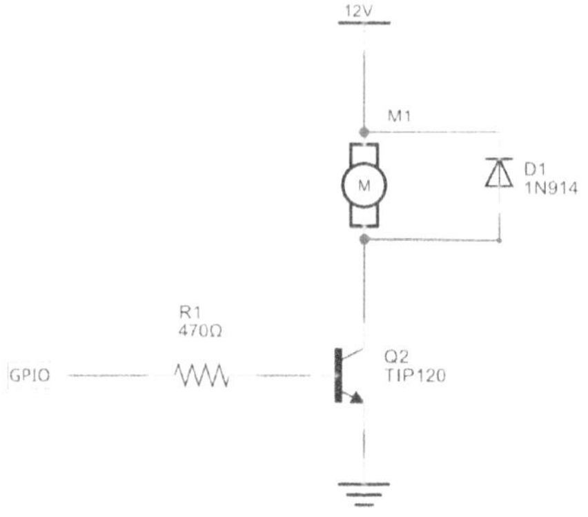

This circuit is simple and will work with motor voltages up to 40V and motor currents up to 5A continuous, 8A peak. The only small point to note is that the TIP120 is a Darlington pair, i.e. it is two transistors in the same case, and as such the base voltage drop is twice the usual 0.6V, i.e. 1.2V, and this has to be taken into account when calculating the current-limiting resistor.

It is sometimes said that the TIP120 and similar are inefficient power controllers because, comprising two transistors, they have twice the emitter-collector voltage you would expect, which means they dissipate more power than necessary.

If you are running a motor from a battery you might want to use a MOSFET, but, as described earlier, 3.3V is low to switch a MOSFET on and off. One solution is to use a BJT to increase the voltage applied to the gate:

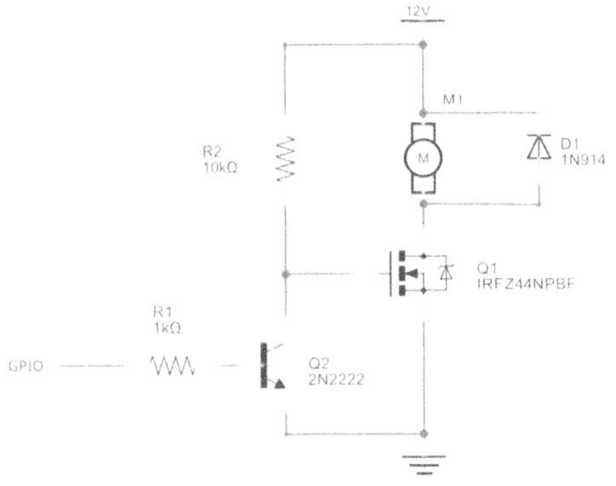

The BJT connects the gate to 12V. As the IRFZ44NPBF has a threshold voltage between 2V and 4V, devices should work at 5V and sometimes at 3.3V without the help of the BJT, but providing 12V ensures that the MOSFET is fully on. One problem with the circuit is that the use of the BJT inverts the signal. When the GPIO line is high the BJT is on and the MOSFET is off and vice versa. In other words, GPIO line high switches the motor off and low switches it on. This MOSFET can work with voltages up to 50V and currents of 40A. The 2N2222 can only work at 30V, or 40V in the case of the 2N2222A.

A third approach to controlling a unidirectional motor is to use half an H-bridge. Why this is so-called, and why you might want to do it, will become apparent in the next section on bidirectional motors. Half an H-bridge makes use of two complementary devices, either an NPN and a PNP BJT or an N- and P-type MOSFET.

For example:

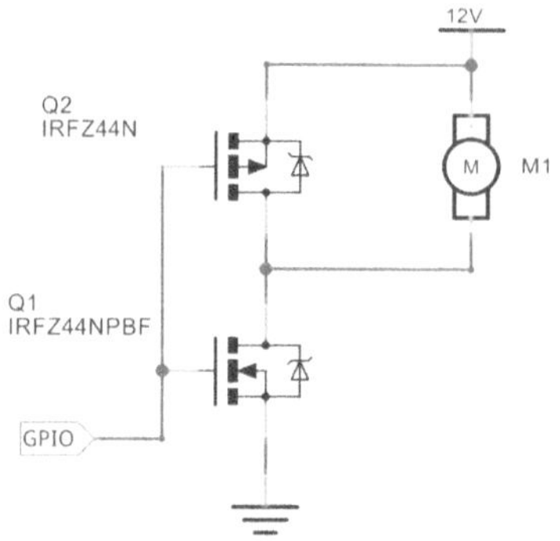

If the GPIO line is high then Q1 is on and Q2 off and the motor runs. If the GPIO line is low then Q1 is off and Q2 is on and the motor is braked – it has a resistance to rotating because of the back electromotive force (EMF) generated when the rotor turns. You probably need a BJT to feed the MOSFETs as selected.

Unidirectional PWM Motor Controller

A function to control the speed of a unidirectional motor is very simple. The speed is set by the duty cycle – the only parameter you have to choose in addition is the frequency. If you want an optimal controller then setting the frequency is a difficult task. Higher speeds make the motor run faster and quieter – but too high a frequency and the motor loses power and the driving transistor or MOSFET becomes hot and less efficient. The determining factor is the inductance of the motor's coil and any other components connected to it such as capacitors. In practice, PWM frequencies from 100Hz to 20kHz are commonly used, but in most cases 1kHz to 2kHz is a good choice.

How should we implement code to make motor control easy? A good pattern is to create an object which has fields that represent the state of the entity and methods to control it. For example, to implement a unidirectional motor we can create a `Motor` class:

```
class Motor:
    def __init__(self, pinNo):
        self.gpio = pinNo
        self._on = False
        self.speed=0

        self.pwm1=PWM(Pin(pinNo))
        self.pwm1.freq(2000)
        self.pwm1.duty_u16(0)
```

You can see that this has all of the information needed to define the current state of a motor. All we need now are some functions to modify the fields and implement the changes to the state.

First we need a function to set the speed:

```
    def setSpeed(self,s):
        self._on=True
        self.speed=s
        self.pwm1.duty_u16(int(65535*s/100))
```

and two functions to turn the motor on and off:

```
    def off(self):
        self._on=False
        self.pwm1.duty_u16(0)
    def on(self):
        self._on=True
        self.pwm1.duty_u16(int(65535*self.speed/100))
```

After this we can create and use a motor very easily. A full program complete with a demonstration is:

```
from machine import Pin, PWM
from time import sleep
class Motor:
    def __init__(self, pinNo):
        self.gpio = pinNo
        self._on = False
        self.speed=0
        self.pwm1=PWM(Pin(pinNo))
        self.pwm1.freq(2000)
        self.pwm1.duty_u16(0)
    def setSpeed(self,s):
        self._on=True
        self.speed=s
        self.pwm1.duty_u16(int(65535*s/100))
    def off(self):
        self._on=False
        self.pwm1.duty_u16(0)
    def on(self):
        self._on=True
        self.pwm1.duty_u16(int(65535*self.speed/100))
motor=Motor(16)
motor.setSpeed(50)
sleep(1)
motor.off()
sleep(1)
motor.setSpeed(90)
sleep(1)
motor.off()
```

This sets up a motor connected to GP16, sets it to 50% speed, pauses, turns it off, then on again at 90% and finally off.

Bidirectional Brushed Motor

If you want bidirectional control then you need to use an H-bridge:

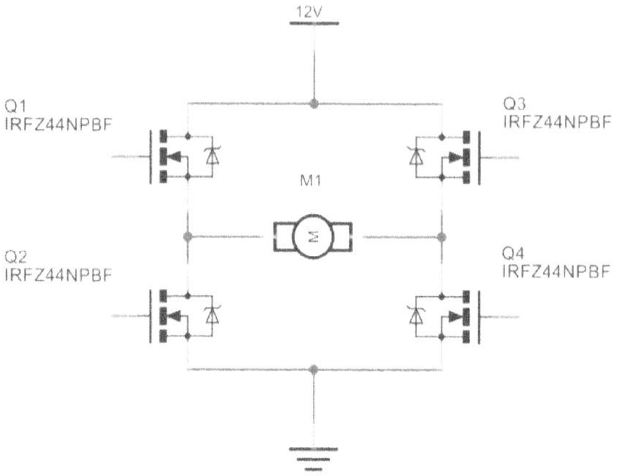

It is easy to see how this works. If Q1 and Q4 are the only MOSFETs on the motor, + is connected to 12V and – to ground. The motor runs in the forward direction. If Q2 and Q3 are the only MOSFETs on the motor, + is connected to ground and – is connected to 12V. The motor runs in the reverse direction. Of course, if none or any single one is on the motor is off. If Q1 and Q3, or Q2 and Q4, are on then the motor is braked as its windings are shorted out and the back EMF acts as a brake.

You can arrange to drive the four MOSFETs using four GPIO lines - just make sure that they switch on and off in the correct order. To make the bridge easier to drive, you can add a NOT gate to each pair so that you switch Q1/Q2 and Q3/ Q4 to opposite states.

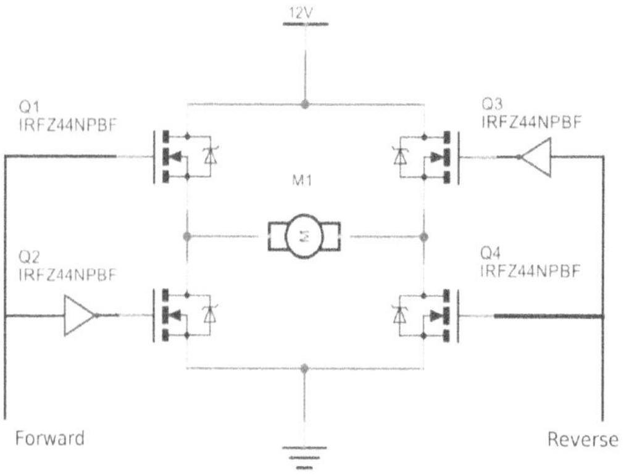

An alternative design is to use complementary MOSFETs:

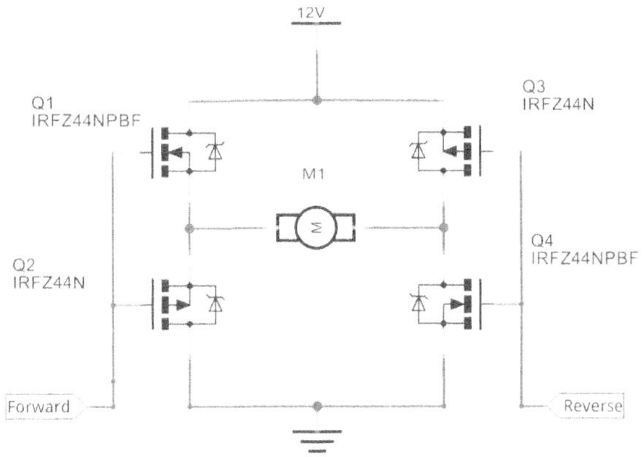

In this configuration, the first GPIO line drives the motor forward and the second drives it in reverse. The effect of setting the two lines is:

Forward	**Reverse**	**Motor**
Low	Low	Off
Low	High	Reverse
High	Low	Forward
High	High	Braked

You can also drive the GPIO lines for Forward/Reverse with a PWM signal and control the motor's speed as well as direction. If you use the MOSFETs shown in the diagram then you would also need a BJT to increase the drive voltage to each MOSFET, as in the unidirectional case. You also need to include diodes to deal with potential reverse voltage on each of the MOSFETs. The most important thing about an H-bridge is that Q1/Q2 and Q3/Q4 should never be on together – this would short circuit the power supply.

If working with four power BJTs or MOSFETs is more than you want to tackle, the good news is that there are chips that implement two H-bridges per device. You can also buy low-cost ready-made modules with one or more H-bridges. One of the most used devices is the L298 Dual H-bridge which works up to 46V and total DC current of 4A.

The block diagram of one of the two H-bridges shows exactly how it works:

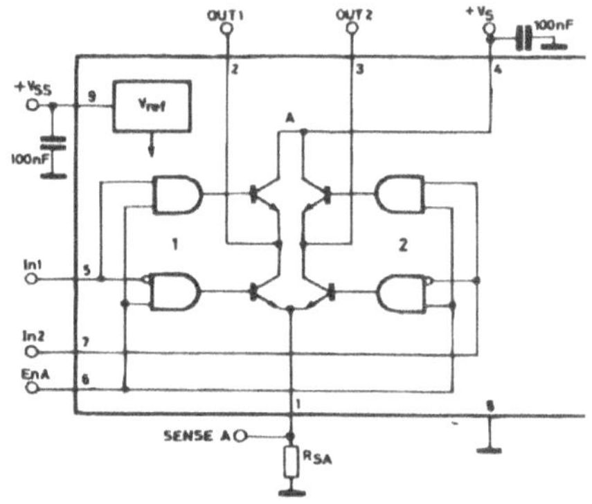

You can see that the bridge is made up of four BJTs and there are logic gates to allow IN1 and IN2 to select the appropriate pairs of devices. The only extras are AND gates and that the ENA (enable) line is used to switch all of the transistors off. The line shown as SENSE A can be used to detect the speed or load of the motor, but is rarely used.

A typical module based on the L298 can be seen below.

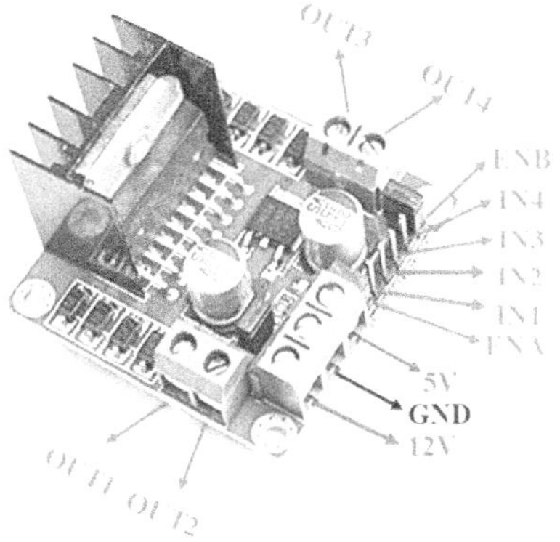

It is easier to describe how to use this sort of module with a single motor. The motor is connected to OUT1 and OUT2. Three GPIO lines are connected to ENA, IN1 and IN2. ENA is an enable line, which has to be high for the motor to run at all. IN1 and IN2 play the role of direction control lines – which one is forward and which is reverse depends on which way round you connect the motor. Putting a PWM signal onto ENA controls the speed of the motor and this allows IN1 and IN2 to be simple digital outputs.

Notice that the power connector shows 5V and 12V supplies, but most of these modules have a voltage regulator which will reduce the 12V to 5V. In this case you don't have to supply a 5V connection. If you want to use more than 12V then the regulator has to be disconnected and you need to arrange for a separate 5V supply – check with the module's documentation. Notice that the transistors in the H-bridge have around a 2V drop, so using 12V results in just 10V being applied to the motor.

Another very popular H-bridge device is the SN754410 driver. This is suitable for smaller, lower-powered, motors and has two complete H-bridges. It can supply up to 1A per driver and work from 4.5 to 36V. It has the same set of control lines as the L298, i.e. each motor has a forward/reverse control line and an enable line. You don't have use the enable line - it can be connected to +5V to allow PWM to be applied on the forward/reverse lines.

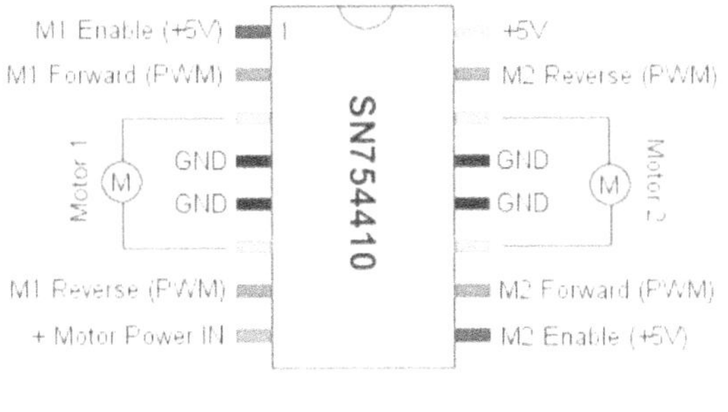

Bidirectional Motor Software

We can easily extend the Motor class and its methods to work with a bidirectional motor. The only real difference is that now we have to use two GPIO lines and these are most easily used if they are the A and B channels of the same slice – note for simplicity the code doesn't check this condition and you specify the GPIO line for channel A in the constructor. For forward we activate the first GPIO line and for reverse we activate the second:

```
class BiMotor(Motor):
    def __init__(self, pinNo):
        super().__init__(pinNo)
        self.forward=True
        self.pwm2=PWM(Pin(pinNo+1))
        self.pwm2.duty_u16(0)

    def setForward(self,forward):
        if self.forward==forward:
            return
        self.pwm1.duty_u16(0)
        self.pwm1,self.pwm2=self.pwm2,self.pwm1
        self.forward=forward
        self.pwm1.duty_u16(int(65535*self.speed/100))

motor=BiMotor(16)
motor.setSpeed(50)
sleep(1)
motor.setForward(False)
sleep(1)
motor.setSpeed(90)
sleep(1)
motor.setForward(True)
motor.off()
```

Notice that changing direction is just a matter of swapping pwm1 and pwm2 – only pwm1 is active and driving the motor.

This is a very basic set of functions, you can add others to improve motor control according to how sophisticated you want it to be. For example, a brake function would set both lines high for brake mode, or you could introduce limits on how fast the speed can be changed.

There are H-bridges that use two lines to control Phase and Enable (PWM). These map to the usual Forward, Reverse and Enable as shown below:

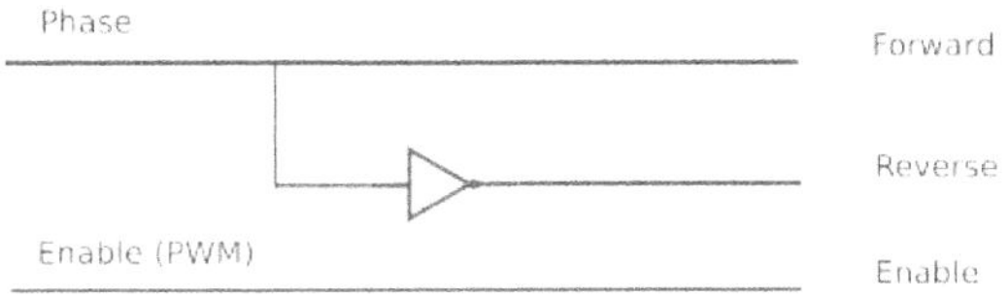

The disadvantage of this arrangement is that you cannot set Forward and Reverse to put the motor into brake mode. If you do have this sort of controller, anything based on the MAX14870/2 for example, then you can modify the functions to use a single PWM line and one standard GPIO line for speed and direction.

Using A Single Full H-Bridge As Two Half H-Bridges

It is easy to think of an H-bridge as being only for bidirectional control, but each full bridge is composed of two half bridges and this means a typical dual full H-bridge can control four unidirectional motors:

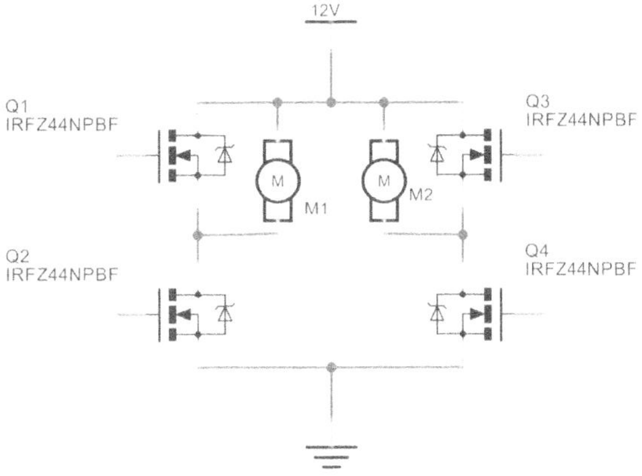

In this case Forward is now MotorM1 speed control and Reverse is now MotorM2 speed control. Any enable line has to be set high to allow the two motors to be controlled. You can make use of this arrangement with the unidirectional software given earlier.

Controlling a Servo

Hobby servos, of the sort used in radio control models, are very cheap and easy to use and they connect via a standard PWM protocol. Servos are not drive motors, but positioning motors. That is, they don't rotate at a set speed, they move to a specified angle or position.

A servo is a motor, usually a brushed DC motor, with a feedback sensor for position, usually a simple variable resistor (potentiometer) connected to the shaft. The output is usually via a set of gears which reduces the rotation rate and increases the torque. The motor turns the gears, and hence the shaft, until the potentiometer reaches the desired setting and hence the shaft has the required angle/position.

A basic servo has just three connections, ground, a power line and a signal line. The colors used vary, but the power line is usually red, ground is usually black or brown and the signal line is white, yellow or orange. If a standard J-connector is fitted then the wire nearest the notch, pin 3, is Signal, the middle wire, pin 2, is 5V and outer wire, pin 1, is Ground.

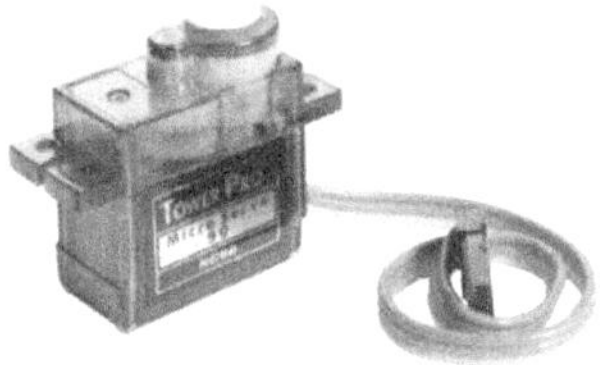

The power wire has to be connected to a 5V supply capable of providing enough current to run the motor - anything up to 500mA or more depending on the servo. The good news is that the servo's signal line generally needs very little current, although it does, in theory, need to be switched between 0V and 5V using a PWM signal.

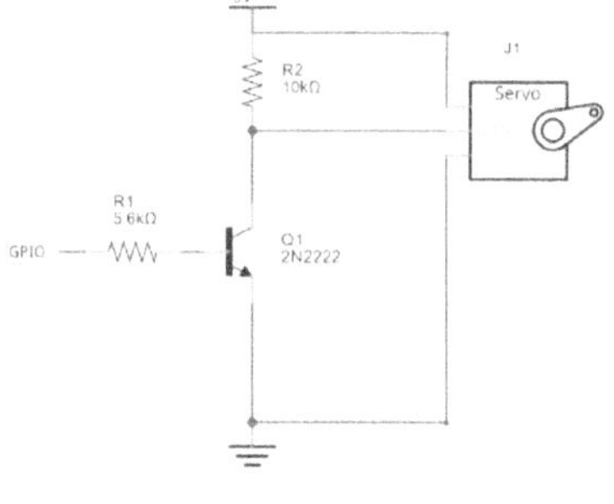

You can assume that the signal line needs to be driven as a voltage load and so the appropriate way to drive the servo is:

- The servo's + line needs to be connected to an external 5V power supply.
- The 10K resistor R1 can be a lot larger for most servos - 47K often works. The 5.6K resistor limits the base current to slightly less than 0.5mA.

Notice, however, that if you are using a single BJT driver, like the one shown above, the input is inverted.

This is the correct way to drive a servo, but in nearly all cases you can drive the servo signal line directly from the 3.3V GPIO line with a 1K resistor to limit the current if anything goes wrong with the servo. Some servos will even work with their motor connected to 3.3V, but at much reduced torque.

Now all we have to do is set the PWM line to produce 20ms pulses with pulse widths ranging from 0.5ms to 2.5ms – i.e. a duty cycle of 2.5 to 12.5%.

Once again, it is easier to use a class to represent the current state of a servo and create methods to change the state:

```
from machine import Pin, PWM
from time import sleep
class Servo:
    def __init__(self, pinNo):
        self.pwm = PWM(Pin(pinNo))
        self.pwm.freq(50)
        self.position = 65535*2.5/100
    def setPosition(self, p):
        self.position = p
        self.pwm.duty_u16(int(65535*p/1000 + 65535*2.5/100))
servo=Servo(16)
servo.setPosition(100.0)
sleep(1)
servo.setPosition(50.0)
sleep(1)
servo.setPosition(0)
```

Notice that we set the frequency to 50Hz. If you want to work with a non-standard servo you can change this or make it settable. The `setPosition` function sets the position in terms of percentages. That is, `setPosition(50)` sets the servo to the middle of its range. This assumes that the servo has a standard positioning range and most don't. In practice, to get the best out of a servo you need to calibrate each servo and discover what range of movement is supported. The main program creates a Servo object on GP16 and then moves the servo to its maximum, middle and minimum positions.

If you run the program using the transistor circuit given earlier, you will discover that the servo does nothing at all, apart perhaps from vibrating. The reason is that the transistor voltage driver is an inverter. When the PWM line is high, the transistor is fully on and the servo's pulse line is effectively grounded. When the PWM line is low, the transistor is fully off and the servo's pulse line is pulled high by the resistor. The solution is to use an inverted output from the GPIO line using the polarity inverting function given in the previous chapter. However, to make this easier to use we need to integrate it into the Servo class:

```
def setPolarity(self,invert):
   sliceNo=self.getSlice()
   channel=self.getChannel()
   Addr = 0x40050000 +0x14*sliceNo
   if invert:
     machine.mem32[Addr]=machine.mem32[Addr] | 0x1 << (2+channel)
   else:
     machine.mem32[Addr]=machine.mem32[Addr] & ~(0x1 << (2+channel))
```

We also need two helper functions that convert the GPIO number to the slice and channel number:

```
def getSlice(self
   return (self.pin>>1) & 0x07

def getChannel(self):
   return self.pin & 1
```

and we need a new attribute to record the pin number:

```
self.pin=pinNo
```

The complete class, and a main program to make use of it inverted, is:

```
from machine import Pin, PWM
import machine
from time import sleep
class Servo:
    def __init__(self, pinNo):
        self.pwm = PWM(Pin(pinNo))
        self.pin=pinNo
        self.pwm.freq(50)
        self.position = 65535*2.5/100

    def setPosition(self, p):
        self.position = p
        self.pwm.duty_u16(int(65535*p/1000 + 65535*2.5/100))
```

```
    def getSlice(self):
        return (self.pin>>1) & 0x07

    def getChannel(self):
        return self.pin & 1


    def setPolarity(self,invert):
        sliceNo=self.getSlice()
        channel=self.getChannel()
        Addr = 0x40050000 +0x14*sliceNo
        if invert:
            machine.mem32[Addr]=machine.mem32[Addr] |
                                    0x1 << (2+channel)
        else:
            machine.mem32[Addr]=machine.mem32[Addr] &
                                    ~(0x1 << (2+channel))


servo=Servo(16)
servo.setPolarity(True)
servo.setPosition(100.0)
sleep(1)
servo.setPosition(50.0)
sleep(1)
servo.setPosition(0)
```

It is worth mentioning that servos make good low-cost DC motors, complete with gearboxes. All you have to do is open the servo, unsolder the motor from the control circuits and solder two wires to the motor. If you want to use the forward/reverse electronics you can remove the end stops on the gearbox, usually on the large gearwheel, and replace the potentiometer with a pair of equal value resistors, 2.2kΩ, say.

Brushless DC Motors

Brushless DC motors are more expensive than brushed DC motors, but they are superior in many ways. They don't fail because of commutator or brush wear and need no maintenance. They provide maximum rotational torque at all points of the rotation and generally provide more power for the same size and weight. They can also be controlled more precisely. The only negative points are higher cost and slightly more complex operation. In practice, it is usually better to use a brushed DC motor unless you really need something extra.

A brushless DC motor is basically a brushed motor turned inside out – the stator is a set of electromagnets and the rotor is a set of permanent magnets. In some designs the permanent magnets are inside the stator in the manner of a brushed motor, an inrunner, and sometimes the magnets are outside of the stator, an outrunner.

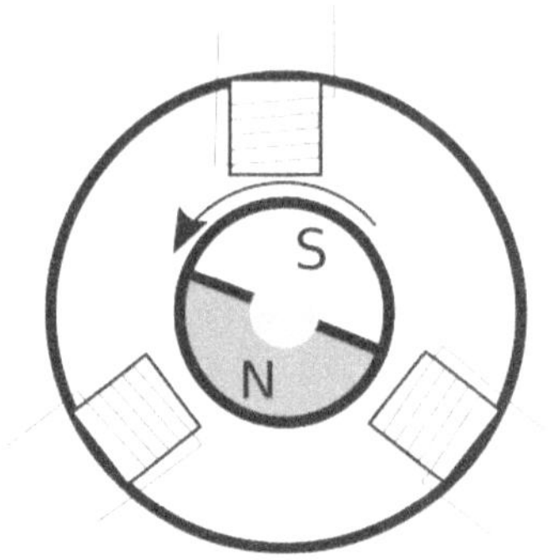

An inrunner – the permanent magnets form the rotor and the coils are switched to attract.

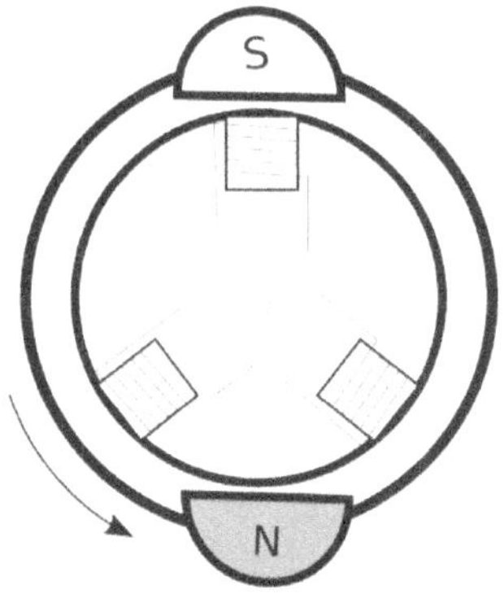

An outrunner – the permanent magnets are on the outside of the stator and the whole cover rotates.

A brushless motor works in exactly the same way as a brushed motor. As the coils are stationary there is no need for a mechanical commutator, but there is still need for commutation – the coils have to be switched on in sequence to create a rotating magnetic field which pulls the rotor around with it. This means that you have to implement an electronic commutator, which is another name for a brushless DC motor.

An electronic commutator has to sense the position of the rotor and change the magnetic field generated by the stator to keep the rotor moving. Brushless motors differ in the number of magnets they have and the number of phases. The most common is a three-phase motor as these are used in radio control modeling. Essentially you need at least a driver for each of the phases and a GPIO line to generate the signal. In practice, you need two drivers for each phase and they have to be driven from a dual supply so that the magnetic field can be positive, zero or negative.

This would be possible to do with software, but it isn't easy and a more reasonable alternative is to buy a ready-built controller. There are two types of brushless motor – with Hall sensors and without. The former are more expensive, but easier to control because the electronics always knows where the rotor is and can apply the correct drive. The ones without sensors are controlled by measuring the back EMF from the motor and this is much harder. Most of the lower-cost speed controllers need motors with Hall sensors.

The radio control community has taken to using three-phase brushless motors and this has resulted in a range of motors and controllers at reasonable prices intended as high-power, high-speed, unidirectional motors for use in quadcopters and model planes.

If you can live with their limitations they provide a good way to couple a Pico to a brushless motor. In this case all you need is a three-phase brushless motor of the sort used in RC modeling and an ESC (Electronic Speed Controller) of the sort shown below:

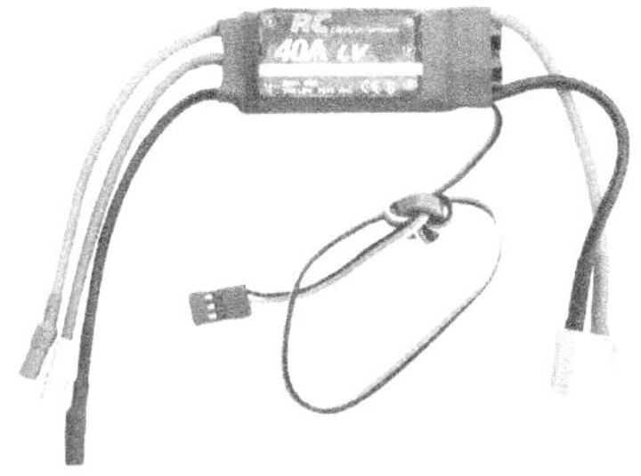

The three leads on the left go to the three phases of the motor and the red and black leads on the right go to a power supply – often a LiPo battery. The small three-wire connector in the middle is a standard servo connector and you can use it exactly as if the brushless motor was a servo, with a few exceptions. The first is that pin 2 supplies 5V rather than accepts it. Don't connect this to anything unless you want a 5V supply. The second problem is that ESCs are intelligent. When you first apply power they beep and can be programmed into different modes by changing the PWM signal from Max to Min. Also, to use an ESC you have to arm it. This is to avoid radio control modelers from being injured by motors that start unexpectedly when the power is applied. The most common arming sequence is for the ESC to beep when power is applied. You then have to set the PWM to Min, when the ESC will beep again. After a few moments you will have control of the motor.

The need for an arming procedure should alert you to the fact that these model motors are very powerful. Don't try working with one loose on the bench as it will move fast if switched on and at the very least make a twisted mess of your wires. Most importantly of all, don't run a motor with anything attached to it until you have everything under control.

Stepper Motors

There is one sort of brushless motor that is easy to use and low cost – the stepper motor. This differs from a standard brushless motor in that it isn't designed for continuous high-speed rotation. A stepper motor has an arrangement of magnets and coils such that powering some of the coils holds the rotor in a particular position. Changing which coils are activated makes the rotor turn until it is aligned with the coils and stops moving. Thus the stepper motor moves the rotor in discrete steps. This makes driving it much simpler, but note it doesn't use PWM for speed control.

Stepper motors have no brushes and so don't wear out as fast as brushed motors. They also have roughly the same torque at any speed and can be used at low speeds without a gearbox. They can remain in a fixed position for a long time without burning out, as DC motors would. Unlike a servo, however, if a stepper motor is mechanically forced to a new position, it will not return to its original position when released. The only disadvantage of a stepper motor is that the continuous rotation produced by repeated stepping can make the motor vibrate.

Stepper motors vary in the size of step they use – typically 1.8 degrees giving 200 steps per rotation, although gearing can be used to reduce the step size. Another big difference is that the rotor is made up either of permanent magnets or soft iron. The first type is called a Permanent Magnet or PM stepper and the second is called a Variable Reluctance or VR and they differ in how you drive them with PM steppers being easier to understand. There are also hybrid steppers which share the good characteristics of both PM and VR stepper motors. These are more expensive and are generally only used where accuracy of positioning is important. They also differ in the number of phases, i.e. independent banks of coils, they have.

The diagram below shows a two-phase PM motor with Phase 1 activated:

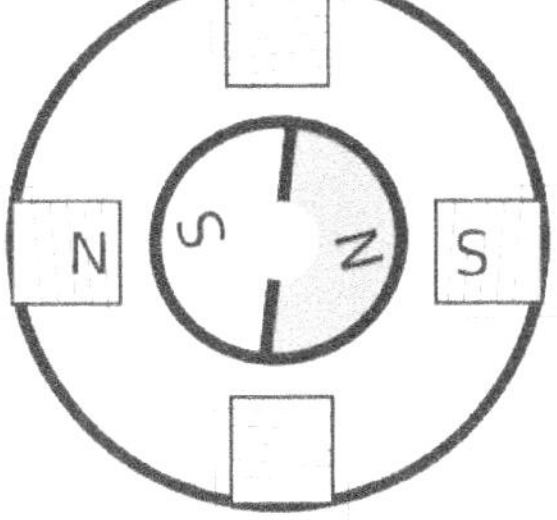

If Phase two is activated, the rotor turns through 90 degrees:

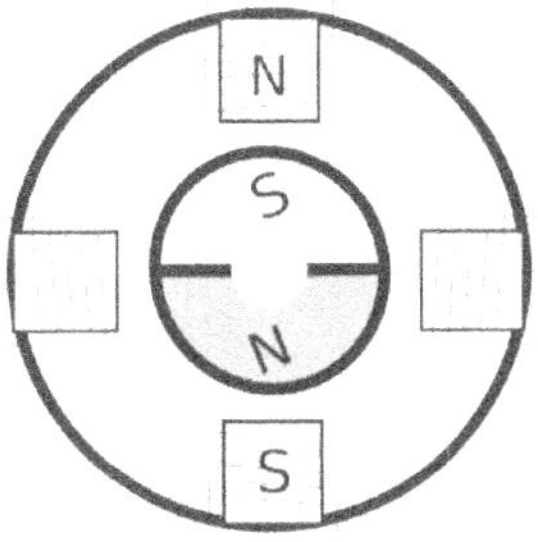

This is the simplest stepper motor you can make. A typical stepper motor will have many more coils than four, but they are usually connected into two or three phases.

Another big difference is bipolar versus unipolar. A bipolar motor is like the one shown in the diagram. To generate a north pole the current has to flow in the opposite direction to when you want to create a south pole. This means you have to drive each bank of coils with a bidirectional driver, e.g. an H-bridge. A unipolar motor has two windings, one in each direction, and both windings can be driven by a unidirectional driver – one giving a north pole and the other a south pole. Notice that a unipolar motor has twice the number of coils to drive and the switching sequence is slightly different.

A two-phase bipolar motor with Phases A and B would switch on in the sequence:

A → B → A- → B- → A etc

where the minus sign means the current flows the other way.

A two-phase unipolar motor has two coils per phase, A1, A2 and B1, B2 with the 1 and 2 windings creating opposite magnetic fields for the same current flow. Now the sequence is:

A1 → B1 → A2 → B2 → A1 etc

and all driven in the same direction.

Switching single phases fully on and off in sequence makes the motor make repeated steps. You can also switch on more than one phase at a time to generate micro-steps. For example, in our two-phase example, switching on two phases makes the rotor settle between the two, so producing a half micro-step:

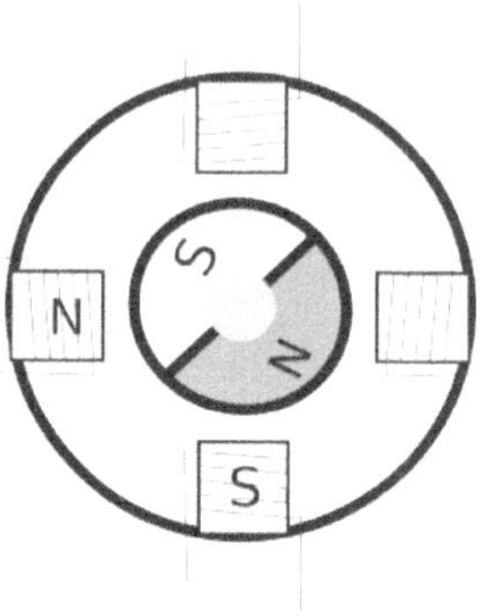

The driving sequence for a two-phase bipolar motor is:

A → AB →B → BA- → A- → A- B- → B- → AB- → A

with minus indicating that the coil is energized in the opposite direction, giving a total of eight, rather than four, steps.

You can even vary the current through the coils and move the rotor to intermediate positions. At the extreme limit you can vary the current sinusoidally and produce a smooth rotation. Micro-stepping is smoother and can eliminate buzzing. For high accuracy positioning, micro-stepping is a poor performer under load.

Stepper Motor Driver

How best to drive a stepper motor using a Pico? There are some specialized chips that work with unipolar and bipolar stepper motors. However, you can easily control a bipolar stepper motor using one the H-bridges described in the section on directional motor control.

For example, using complementary MOSFETS:

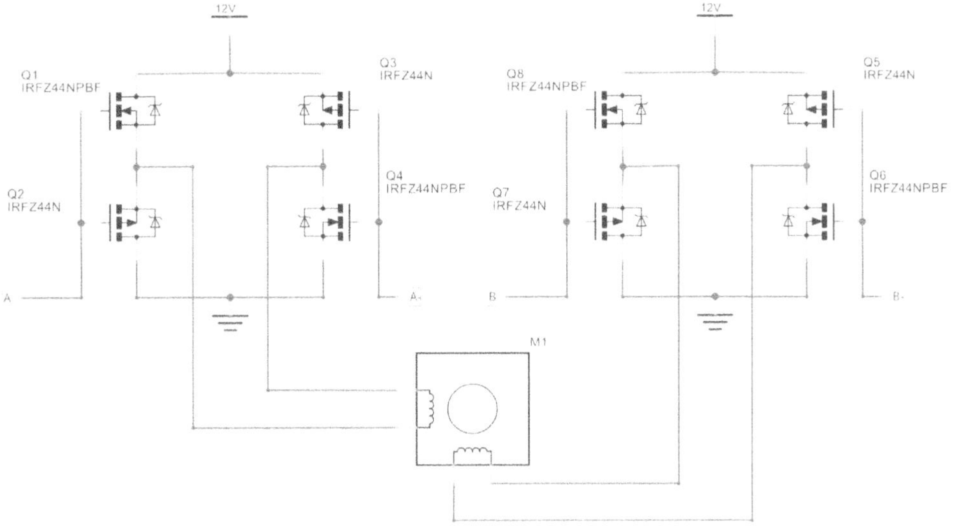

You can use a dual H-bridge module in the same way if you don't want to build it from scratch. The motor has to be a bipolar two-phase motor, often called a four-wire stepper motor. You can see that for this arrangement you need four GPIO lines, A, A-, B and B-.

What about driving the dual H-bridge using software? You need four GPIO lines and you need to pulse them in a specific phase to make the motor rotate. This could be done using the PIO, see Chapter 13, but it is easier to start with software-controlled GPIO lines. The first question to answer is how to specify the four GPIO lines to be used. As we will see, there is a big advantage and simplification in using a block of four consecutive lines. The reason is that we can easily set the bits corresponding to these lines in a mask and so set them in a single operation. In this case we simply record the number of the numerically lowest GPIO line GPn and assume that a block of four GPIO lines are allocated GPn → A, GPn+1 → B, GPn+2 → A- and GPn+3 → B-.

As before, we use a class to record the current state of the motor and its methods:

```
class StepperBi4():
    def __init__(self, pinA):
        self.phase=0
        self.pinA=pinA
        self.gpios= tuple([Pin(pinA,Pin.OUT),Pin(pinA+1,Pin.OUT),
                           Pin(pinA+2,Pin.OUT),Pin(pinA+3,Pin.OUT)])
        self.gpios[0].high()
        self.gpioMask=0xF<<self.pinA
        self.halfstepSeq =[0x1, 0x3, 0x2, 0x6, 0x4, 0xC, 0x8, 0x9]
#   [
#             [0,0,0,1],
#             [0,0,1,1],
#             [0,0,1,0],
#             [0,1,1,0],
#             [0,1,0,0],
#             [1,1,0,0],
#             [1,0,0,0],
#             [1,0,0,1]
#    ]

    def _gpio_set(self,value,mask):
        machine.mem32[0xd0000000+0x01C]=
                (machine.mem32[0xd0000000+0x010]^value) & mask

    def setPhase(self,phase):
        value=self.halfstepSeq[phase] << self.pinA
        self._gpio_set(value,self.gpioMask)
        self.phase=phase

    def stepForward(self):
        self.phase=(self.phase+1) % 8
        self.setPhase(self.phase)

    def stepReverse(self):
        self.phase=(self.phase-1) % 8
        self.setPhase(self.phase)


step=StepperBi4(16)
step.setPhase(0)

while True:
    step.stepForward()
    sleep_ms(1)
```

You can see that the List halfstepSeq is just the hex value corresponding to the binary representation of which line is high and which is low in each state. The bits correspond to the GPIO lines such that the low order bits control the lowest numbered GPIO line.

You can see that if GP16 is A, GP17 is B, GP18 is A- and GP19 is B- and stepping through the `stepTable` gives the sequence given earlier:

A → AB → B → BA- → A- → A-B- → B- → AB- → A

Now all we need are some methods to set and work with the phase:

The `setPhase` function uses the `_gpio_set` function given earlier to change only the GPIO lines we are using to the bit pattern in the `stepTable` for the specified phase. The `stepForward` and `stepReverse` simply move down or up in the phase table, making sure to go back to the start when the end is reached. This is what the modulus operator, `%`, does for us and `phase` follows the sequence `0,1,2,3,4,5,6,7,0,1` and so on.

You can use a full step table if you want to as long as you remember to work in mod 4 rather than 8:

```
self.stepSeq =[0x1,0x2,0x4,0x8]
# [
#           [0, 0, 0, 1],
#           [0, 0, 1, 0],
#           [0, 1, 0, 0],
#           [1, 0, 0, 0],
#       ]
#
```

Of course, these are identical to the even elements of the half step table so we can achieve the same result by using that table but increasing the increment by 2 instead of 1:

```
    def stepForward(self):
        self.phase=(self.phase+2) % 8
        self.setPhase(self.phase)

    def stepReverse(self):
        self.phase=(self.phase-2) % 8
        self.setPhase(self.phase)
```

To try either version of the program you need an H-bridge connected so that GP16 is A, GP17 is B, GP18 is A- and GP19 is B-. The maximum stepping speed, i.e. with no timer delay, is such that a 200-step motor, i.e. 400 half steps, will rotate in 400*0.1 ms= 0.04 s, i.e. 1500rpm, which is maximum most stepper motors can run at. In practice include delays to slow things down to a few hundred rpm.

If you are using one of the many dual H-bridge modules then the wiring is as shown below:

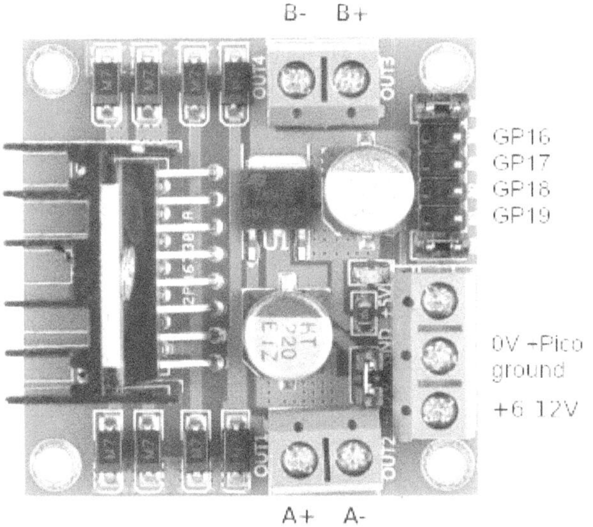

Notice that you have to connect the ground of the power supply and the Pico's ground together. It is also a good idea to use a power supply with a current trip when first trying things out.

The logic analyzer reveals that the switching times are accurate and there is no measurable overlap between pulses:

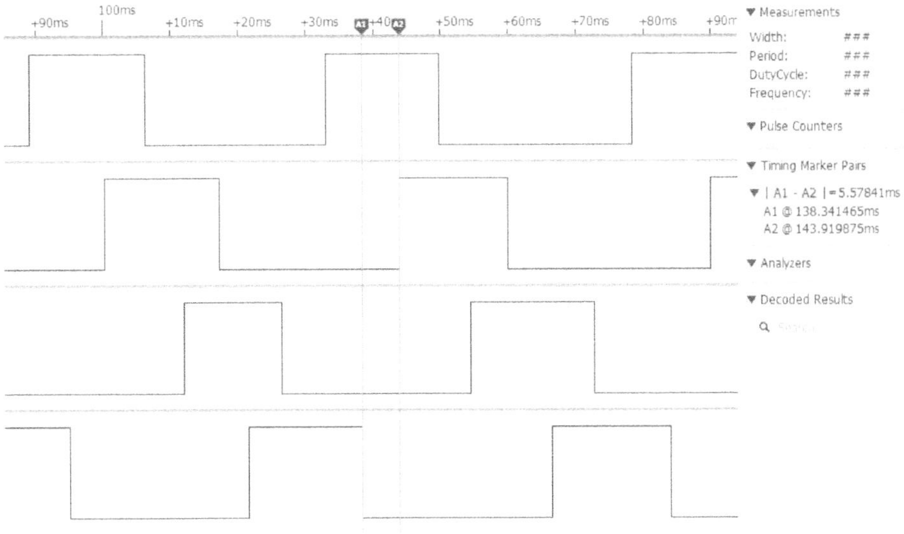

Stepper Motor Rotation – Using Timers

Most of the time you use a stepper motor to move to a given position by executing an exact number of steps. If you do want to make a stepper motor rotate then you need to arrange to step at a regular rate. The simplest way to do this with a Pico is to use a timer interrupt.

To create a Timer object use its constructor:

```
timer=Timer()
```

Once you have a Timer object you can use its init method to set up either a periodic callback or a one-off callback – a one-shot:

```
Timer.init(mode=mode, period=- 1, callback=None)
```

Mode can be either

`Timer.ONE_SHOT`	The timer runs once and the callback is called when the period is up.
`Timer.PERIODIC`	The timer runs repeatedly calling the callback each time the period is up.

When you are finished with a timer you can use:

```
Timer.deinit()
```

to free the hardware and stop any periodic callback.

Notice that the callback function is an interrupt handler and is subject to all of the restrictions and cautions that apply to a general interrupt handler. The callback accepts a single parameter which is the Timer object which called it,

To make the motor rotate on its own we need a timer interrupt handler:

```
    def _doRotate(self,timer):
        if self.forward:
            self.stepForward()
        else:
            self.stepReverse()
```

You can see that this function simply steps the motor forward or backward depending on the setting of the `forward` attribute.

To make this work we need a function to set up the timer interrupt:

```
    def rotate(self,forward,speed):
        self.forward=forward
        self.speed=speed
        if speed==0:
            self.timer.deinit()
            self.timer=None
            return
        if self.timer==None:
            self.timer=machine.Timer()
        self.timer.init(freq=speed,mode=
                     machine.Timer.PERIODIC,callback=self.doRotate)
```

The speed is specified in half steps per second. So setting speed to 200 gives a step rate of 200 half steps per second i.e. 5ms per step or a rotation of 30rpm. Notice that we have to deal with a speed of zero differently by removing the Timer. We also has to check that there isn't already a Timer allocated – if there is we just use it. We also need some additional attributes:

```
        self.forward=True
        self.speed=0
```

A main program to rotate the motor for 500ms at 10ms per step i.e. 0.25pm and then stop it for another 500ms is:

```
step=StepperBi4(16)
step.setPhase(0)
while True:
    step.rotate(True,100)
    sleep_ms(500)
    step.rotate(True,0)
    sleep_ms(500)
```

Notice that when the motor is stopped the phase is still active and so the motor is holds its position under power:

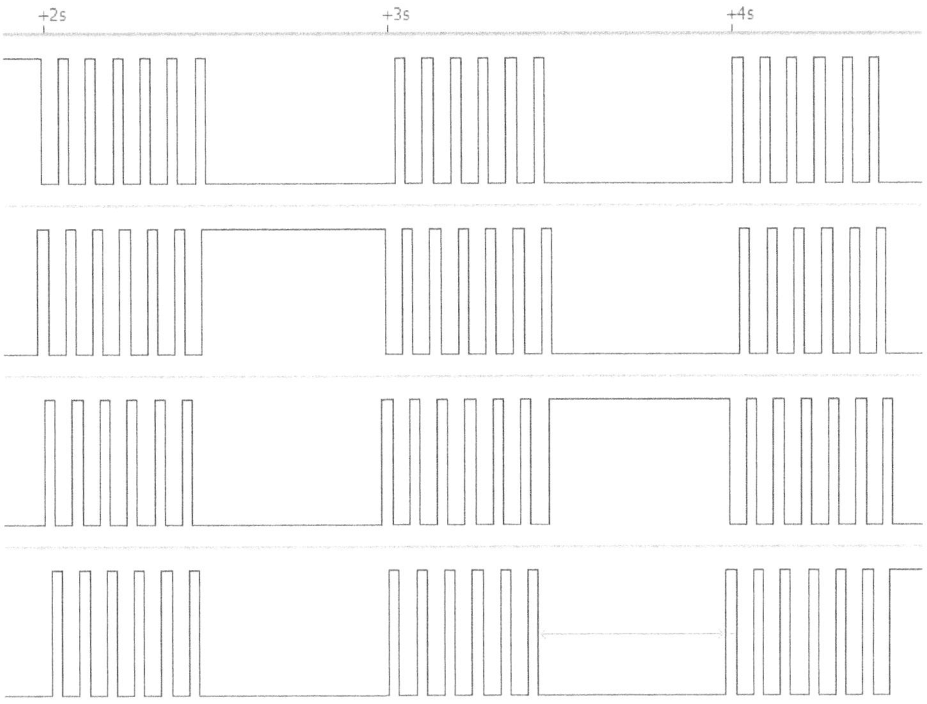

You can elaborate on this basic scheme to include more complex controls on the motor's behavior, including things like allowing it to freewheel and selecting full or half stepping.

The complete program including the timer interrupt handling is:

```
from utime import sleep_ms
from machine import Pin
import machine

class StepperBi4():
    def __init__(self, pinA):
        self.phase=0
        self.pinA=pinA
        self.timer=None
        self.forward=True
        self.speed=0

        self.gpios= tuple([Pin(pinA,Pin.OUT),Pin(pinA+1,Pin.OUT),
                          Pin(pinA+2,Pin.OUT),Pin(pinA+3,Pin.OUT)])
        self.gpios[0].high()
        self.gpioMask=0xF<<self.pinA
        self.halfstepSeq =[0x1, 0x3, 0x2, 0x6, 0x4, 0xC, 0x8, 0x9]
#    [
#              [0,0,0,1],
#              [0,0,1,1],
#              [0,0,1,0],
#              [0,1,1,0],
#              [0,1,0,0],
#              [1,1,0,0],
#              [1,0,0,0],
#              [1,0,0,1]
#     ]

    def _gpio_set(self,value,mask):
        machine.mem32[0xd0000000+0x01C]=
            (machine.mem32[0xd0000000+0x010]^value) & mask

    def setPhase(self,phase):
        value=self.halfstepSeq[phase] << self.pinA
        self._gpio_set(value,self.gpioMask)
        self.phase=phase

    def stepForward(self):
        self.phase=(self.phase+1) % 8
        self.setPhase(self.phase)

    def stepReverse(self):
        self.phase=(self.phase-1) % 8
        self.setPhase(self.phase)
```

```
    def doRotate(self,timer):
        if self.forward:
            self.stepForward()
        else:
            self.stepReverse()

    def rotate(self,forward,speed):
        self.forward=forward
        self.speed=speed
        if speed==0:
            self.timer.deinit()
            self.timer=None
            return
        if self.timer==None:
            self.timer=machine.Timer()
        self.timer.init(freq=speed,mode=
             machine.Timer.PERIODIC,callback=self.doRotate)


step=StepperBi4(16)
step.setPhase(0)
while True:
    step.rotate(True,100)
    sleep_ms(500)
    step.rotate(True,0)
    sleep_ms(500)
```

Summary

- There are a number of different types of electric motor, but DC brushed or brushless motors are the most used in the IoT.
- Brushed motors can be speed controlled using a single transistor driver and a PWM signal.
- For bidirectional control you need an H-bridge. In this case you need two PWM signals.
- Servo motors set their position in response to the duty cycle of a PWM signal.
- Brushless DC motors are very powerful and best controlled using off-the-shelf electronic modules. They are very powerful and thus dangerous if used incorrectly. They can be driven using a simple PWM signal.
- Stepper motors are a special case of a Brushless DC motor. They move in discrete steps in response to energizing different coils.
- A unipolar motor has coils that can be driven in the same direction for every step. A bipolar motor has coils that need to be driven in reverse for some steps.
- Bipolar motors need two H-bridges to operate and four GPIO lines.
- You can easily create a stepper motor driver using four GPIO lines.

Chapter 10

Getting Started With The SPI Bus

The Serial Peripheral Interface (SPI) bus can be something of a problem because it doesn't have a well-defined standard that every device conforms to. Even so, if you only want to work with one specific device it is usually easy to find a configuration that works - as long as you understand what the possibilities are.

SPI Bus Basics

The SPI bus is commonly encountered as it is used to connect all sorts of devices from LCD displays, through realtime clocks and A-to-D converters, but as different companies have implemented it in different ways, you have to work harder to implement it in any particular case. However, it does usually work, which is a surprise for a bus with no standard, or clear, specification.

The reason it can be made to work is that you can specify a range of different operating modes, frequencies and polarities. This makes the bus slightly more complicated to use, but generally it is a matter of looking up how the device you are trying to work with implements the SPI bus and then getting the Pico to work in the same way.

The SPI bus is odd in another way - it does not use bidirectional serial connections. There is a data line for the data to go from the master to the slave and a separate data line from the slave back to the master. That is, instead of a single data line that changes its transfer direction, there is one for data going out and one for data coming in. It is also worth knowing that the drive on the SPI bus is push-pull and not open-collector/drain. This provides higher speed and more noise protection as the bus is driven in both directions. There is a bidirectional mode, where a single wire is used for the data, and the Pico supports this.

In the configuration most used for the Pico, there is a single master and, at most, two slaves. The signal lines are:

- MOSI (Master Output Slave Input), i.e. data to the slave
- MISO (Master Input Slave Output), i.e. data to the master
- SCLK (Serial Clock), which is always generated by the master

In general, there can also be any number of SS (Slave Select), CE (Chip Enable) or CS (Chip Select) lines, which are usually set low to select which slave is being addressed. Notice that unlike other buses, I2C for example, there are no SPI commands or addresses, only bytes of data. However, slave devices do interpret some of the data as commands to do something or send some particular data.

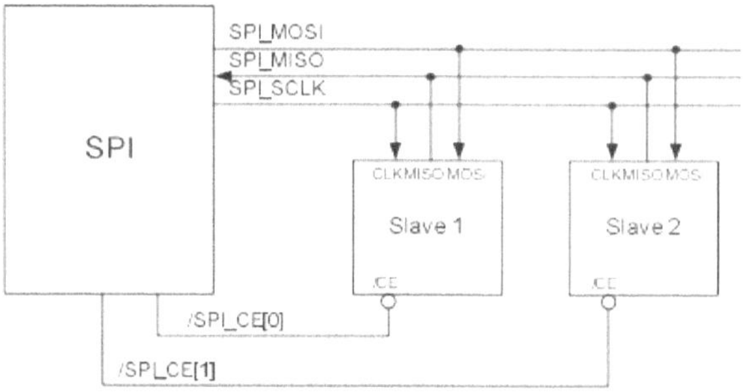

There are two other modes of operation of the SPI interface – bidirectional and LoSSI mode. The bidirectional mode simply uses a single data line, MIMO, for both input and output. The direction of the line is determined by writing a command to the slave. LoSSI mode is used to communicate with sophisticated peripherals such as LCD panels. Both of these are unsupported and beyond the scope of this chapter. However, once you know how standard mode works, the other two are simple variations.

The data transfer on the SPI bus is also slightly odd. What happens is that the master pulls one of the chip selects low, which activates a slave. Then the master toggles the clock SCLK and both the master and the slave send a single bit on their respective data lines. After eight clock pulses, a byte has been transferred from the master to the slave and from the slave to the master. You can think of this as being implemented as a circular buffer, although it doesn't have to be.

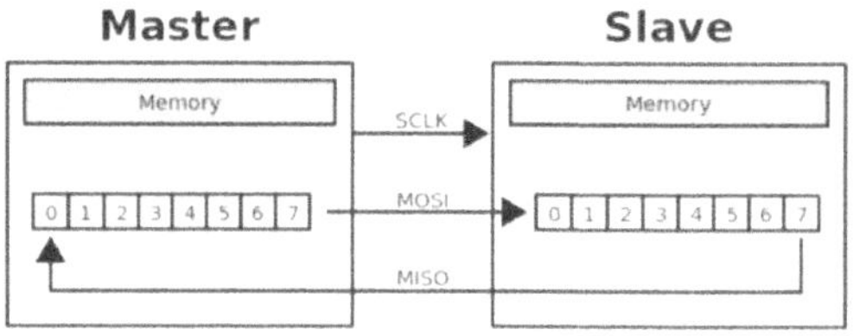

This full-duplex data transfer is often hidden by the software and the protocol used. For example, there is a read function that reads data from the slave and sends zeros or data that is ignored by the slave. Similarly, there is a write function that sends valid data, but ignores whatever the slave sends. The transfer is typically in groups of eight bits, usually most significant bit first, but this isn't always the case. In general, as long as the master supplies clock pulses, data is transferred.

Notice this circular buffer arrangement allows for slaves to be daisy-chained with the output of one going to the input of the next. This makes the entire chain one big circular shift register. This can make it possible to have multiple devices with only a single chip select, but it also means any commands sent to the slaves are received by each one in turn. For example, you could send a convert command to each A-to-D converter in turn and receive back results from each one.

The final odd thing about the SPI bus is that there are four modes which define the relationship between the data timing and the clock pulse. The clock can be either active high or low, which is referred to as clock polarity (CPOL), and data can be sampled on the rising or falling edge of the clock, which is clock phase (CPHA).

All combinations of these two possibilities gives the four modes:

SPI Mode*	Clock Polarity CPOL	Clock Phase CPHA	Characteristics
0	0	0	Clock active high data output on falling edge and sampled on rising
1	0	1	Clock active high data output on rising edge and sampled on falling
2	1	0	Clock active low data output on falling edge and sampled on rising
3	1	1	Clock active low data output on rising edge and sampled on falling

*The way that the SPI modes are labeled is common but not universal.

There is often a problem trying to work out what mode a slave device uses. The clock polarity is usually easy and the Clock phase can sometimes be worked out from the data transfer timing diagrams and:

- First clock transition in the middle of a data bit means CPHA=0
- First clock transition at the start of a data bit means CPHA=1

So to configure the SPI bus to work with a particular slave device:

1. Select the clock frequency - anything from 125MHz to 3.8kHz
2. Determine the CS polarity - active high or low
3. Set the clock mode Mode0 thru Mode3

Now we have to find out how to do this using MicroPython.

Pico SPI Interfaces

The Pico has two SPI controllers, SPI0 and SPI1, that can work as a master or a slave. The connections from both controllers can be routed to different pins:

SPI0	MSIO	GP0	GP4	GP16	GP20
	CSn	GP1	GP5	GP17	GP21
	SCLK	GP2	GP6	GP18	GP22
	MOSI	GP3	GP7	GP19	GP23
SPI1	MSIO	GP8	GP12	GP24	GP28
	CSn	GP9	GP13	GP25	GP29
	SCLK	GP10	GP14	GP26	
	MOSI	GP11	GP15	GP27	

For a full SPI interface you need to one pin for MSIO, one for MOSI and one for SCLK. The CSn lines are not really part of the SPI implementation. To make use of them you have to treat them like standard GPIO lines and set them high and low to select the device under program control. What this means is that you can use as many of the CSn lines with any SPI interface as you need.

The SPI Functions

Before you can make use of the SPI methods you have to import the SPI module:

```
from machine import SPI
```

and you have to create an SPI object:

```
spi=SPI(n)
```

where n is zero or one to select one of the two SPI controllers. You can also include parameters that are used in the initialization method to the constructor. If you don't the SPI controller keeps the last configuration it was used in.

The SPI methods can be grouped into initialization and data transfer. Let's look at each group in turn.

Initialization

There is an `init` method that can be used to set up the hardware

```
SPI.init(baudrate=1000000, polarity=0, phase=0, bits=8,
          firstbit=SPI.MSB, sck=None, mosi=None, miso=None)
```

At the time of writing this doesn't seem to work as a way to specify the pins in use. Until this is changed you should use:

```
spi=SPI(0, sck=Pin1, mosi=Pin2, miso=Pin3)
spi.init(baudrate=1000000, polarity=0, phase=0, bits=8,
                          firstbit=SPI.MSB)
```

This lets you set the number of data bits, and the SPI mode. Notice that, for the Pico the bit order can only be `SPI_MSB`. The mode is set by specifying polarity and phase using:

`phase=0`	Data latched on first clock transition
`phase=1`	Data latched on second clock transition

together with:

`polarity=0`	Clock active high
`polarity=1`	Clock active low

You have to specify the pins used for `sck`, `mosi` and `miso` in the constructor and these have to be `Pin` objects, not just GPIO numbers.

You can use the:

```
SPI.deinit()
```

function to turn off the SPI buss you have been using.

There is also a software implementation of the SPI protocol which works in the same way as the hardware and can use any GPIO line for any SPI pin. At the time of writing this doesn't seem to work.

Data Transfer Functions

Because of the way the SPI bus uses a full-duplex transfer, things are a little different from other buses when it comes to implementing functions to transfer data. The most basic transfer function is:

```
SPI.write_readinto(write_buf, read_buf)
```

The buffers are byte arrays and they have to be of the same length. The `write_buf` is sent byte-by-byte as data is read into the `read_buf` at the same time. If you don't understand this idea of a send (source) and a receive (destination) byte array then see the discussion about how SPI works given earlier.

Notice that a call to any of the transfer functions is blocking, i.e. the function doesn't return until the transfer is complete. There is also no need for a timeout as all SPI transfers end after the number of clock pulses needed to transfer the specified data. Also all of the transfer functions return the number of elements read/written.

If you are only interesting in reading or writing data then you can use:

```
SPI.read(nbytes, write=0)
```

```
SPI.readinto(buf, write=0)
```

The read function returns a byte array with `nbytes` elements. The `readinto` function uses the specified buffer and reads data until it is full. Both send the byte specified by `write` as the data to the slave. Finally you can send data using:

```
SPI.write(buf)
```

In the case of the write function, the received data is simply discarded.

There is one final complication that you can mostly ignore. The SPI hardware is buffered. Both the MISO and MOSI connections have an 8-deep FIFO buffer and this means that you can send and receive data faster if the buffer isn't full or empty. Unfortunately, MicroPython doesn't provide any way to find the state of the buffer.

Using the Data Transfer Functions

When you first start using SPI it can be difficult to get used to the idea that you send and receive data both at the same time. Indeed, you cannot receive data unless you send the same number of data elements.

The most basic of the transfer functions is `spi_write16_read16_blocking` and it makes the bidirectional transfer obvious. For example:

```
read=bytearray(3)
write=bytearray('ABC','utf-8')
spi.write_readinto(write,read)
```

This sends the three elements in `write` and receives back three elements in `read`. Whether the three elements in `read` make any sense is a matter of what the slave sends back and they may be of no interest at all. If this is the case, you might as well use:

```
write=bytearray('ABC','utf-8')
spi.write(write)
```

Whatever three elements the slave sends back are simply ignored.

In the same way, if what you transmit to the slave isn't of any interest, you could use:

```
read=bytearray(3)
SPI.readinto(read, write=0)
```

This reads three bytes into `read` while sending three null bytes to the slave. Alternatively you could use:

```
read=spi.read(3, write=0)
```

which returns a thee-element byte array while sending null bytes to the slave.

Now we come to a subtle point. What is the difference between transferring multiple bytes and simply sending the bytes individually using multiple transfer calls? The answer is that each time you make a transfer call the chip select line should be activated, the data transferred and then deactivated. Using the buffer transfers, the chip select can be left active for the entire transfer, i.e. it isn't necessary to deactivate it between each byte. Sometimes this difference isn't important and you can transfer three bytes using three calls, or just one, to a transfer function. However, some slaves will abort the current multibyte operation if the chip select line is deactivated in the middle of a multibyte transfer. As the Pico requires you to activate and deactivate the chip select line manually, it is up to you to make the distinction between single and multibyte transfers.

It is important to realize that the nature of the transfer is that the first element is sent at the same time that the first element is received. That is, unlike other protocols, the whole of the send buffer isn't sent before the received data comes back. The entire transfer works a data element at a time – the first element is sent while the first element is being received, then the second element is sent at the same time as the second element is being received and so on. Not fully understanding this idea can lead to some interesting bugs.

A Loopback Example

Because of the way that data is transferred on the SPI bus, it is very easy to test that everything is working without having to add any components. All you have to do is connect MOSI to MISO so that anything sent is also received in a loopback mode.

First we have to select which pins to use and, as this is fairly arbitrary at this stage, we might as well use:

```
SPI0  MSIO   GP4  pin 6
      SCLK   GP6  pin 9
      MOSI   GP7  pin 10
```

We can ignore the CS line at the moment as it isn't used in a loopback.

First, connect pin 6 to pin 10 using a jumper wire and start a new project.

The program is very simple. First we initialize the SPI bus to use SPI0 and pins GP6, GP7 and GP4.

```
spi=SPI(0,sck=Pin(6),miso=Pin(4),mosi=Pin(7) )
```

Next we configure the interface:

```
spi.init(baudrate=500_000,bits=8, polarity=0,
phase=0,firstbit=SPI.MSB )
```

We are using 8-bit data and SPI mode 0.
Check that the received data matches the sent data:

```
read=bytearray(3)
write=bytearray([0xAA,0xAA,0xAA])
spi.write_readinto(write,read)

print(read,write)
```

The hex value `AA` is useful in testing because it generates the bit sequence `10101010` which is easy to see on a logic analyzer. Finally we close the bus:

```
spi.deinit()
```

Putting all of this together gives us the complete program:

```
from machine import Pin, SPI

spi=SPI(0,sck=Pin(6),miso=Pin(4),mosi=Pin(7))
spi.init(baudrate=500_000,bits=8, polarity=0,
                        phase=0,firstbit=SPI.MSB )

read=bytearray(3)
write=bytearray([0xAA,0xAA,0xAA])
spi.write_readinto(write,read)

print(read,write)
spi.deinit()
```

If you run the program and don't get any data received back then the most likely reason is that you have connected the wrong two pins or not connected them at all.

If you connect a logic analyzer to the three pins involved – 4, 6 and 7, you will see the data transfer:

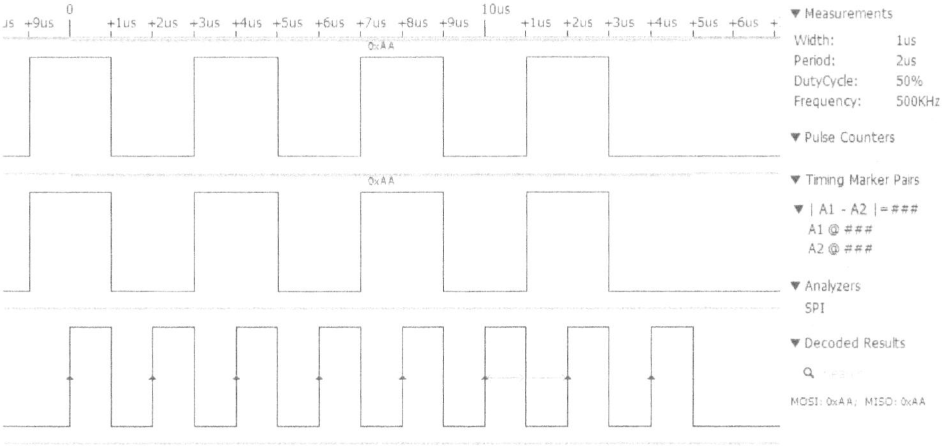

Using The CS Line

Of course there is no CS line activity on the logic analyzer because, unlike many SPI implementations, the Pico doesn't automatically control the CS line. It is up to you to initialize a CS line as a standard GPIO line and set it as appropriate during the data transfer.

For example, let's add GP5:

```
SPI0   MSIO    GP4   pin 6
       CSn     GP5   pin 7
       SCLK    GP6   pin 9
       MOSI    GP7   pin 10
```

Notice that there is no real reason to use GP5, any of the possible CSn pins would do.

Using CS active low we first have to initialize the GPIO line to output and set it high:

```
CS=Pin(5,Pin.OUT)
CS.high()
```

Now we can use the CS line to activate the slave by setting it low before starting the transfer and then setting it back high again afterwards:

```
CS.low()
spi.write_readinto(write,read)
CS.high()
```

If you try this out you might notice that there is a problem if you examine what is happening. The setting of the CS line low occurs about 45μs before the first data or clock bit and the setting back to low occurs around 8μs after the final clock pulse.

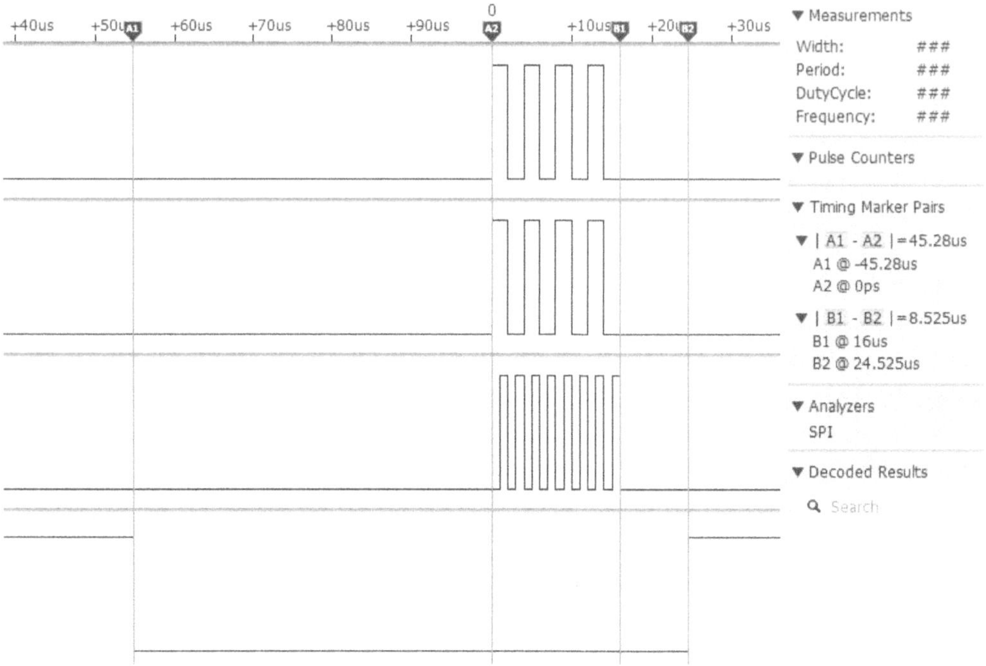

For many devices this over-long CS pulse doesn't make any difference, but it slows down how fast you can send data using the Bus.

The BME280 Humidity, Pressure and Temperature Sensor

As an example of getting to grips with a real SPI device, we can write some code to get data from the commonly used BME280 sensor. You can also use the earlier BMP280, which is register-compatible but doesn't have a humidity sensor. Both work as an SPI and an I2C device and some of their operations as an SPI device are compromised relative to their performance as an I2C device. The BME280 is also the subject of a complete example in the Pico's SDK which implements a full interface, including the correction of raw data values. In this section we will simply implement a basic read of the raw data to show how it is done. If you want to use the device in a real application you will need the additional functions that you can find in the C SDK example they are not difficult to convert to MicroPython.

To make the SDK example easier to use we will connect the device using the same GPIO pins:

PICO	BME280
GP16 (pin 21) MISO/`spi0_rx`	SDO/SDO
GP17 (pin 22) Chip Select	CSB/!CS
GP18 (pin 24) SCK/`spi0_sclk`	SCL/SCK
GP19 (pin 25) MOSI/`spi0_tx`	SDA/SDI
3.3v (pin 36)	VCC
GND (pin 38)	GND

Notice that what the pins are called on a typical BME280 board varies, but you should be able to identify them.

Once you have the device connected what you need to know to send and receive data via the SPI bus varies according to the device and you have little choice but to find and read the datasheet. In this case the important table is the "memory map":

Register Name	Address	bit7	bit6	bit5	bit4	bit3	bit2	bit1	bit0	Reset state
hum_lsb	0xFE	hum_lsb<7:0>								0x00
hum_msb	0xFD	hum_msb<7:0>								0x80
temp_xlsb	0xFC	temp_xlsb<7:4>				0	0	0	0	0x00
temp_lsb	0xFB	temp_lsb<7:0>								0x00
temp_msb	0xFA	temp_msb<7:0>								0x80
press_xlsb	0xF9	press_xlsb<7:4>				0	0	0	0	0x00
press_lsb	0xF8	press_lsb<7:0>								0x00
press_msb	0xF7	press_msb<7:0>								0x80
config	0xF5	t_sb[2:0]			filter[2:0]				spi3w_en[0]	0x00
ctrl_meas	0xF4	osrs_t[2:0]			osrs_p[2:0]			mode[1:0]		0x00
status	0xF3					measuring[0]			im_update[0]	0x00
ctrl_hum	0xF2						osrs_h[2:0]			0x00
calib26..calib41	0xE1...0xF0	calibration data								individual
reset	0xE0	reset[7:0]								0x00
id	0xD0	chip_id[7:0]								0x60
calib00..calib25	0x88...0xA1	calibration data								individual

Registers:	Reserved registers	Calibration data	Control registers	Data registers	Status registers	Chip ID	Reset
Type:	do not change	read only	read / write	read only	read only	read only	write only

The important registers for our simple implementation are the first eight data registers, which give the raw values of humidity, temperature and pressure. These are all read-only registers and the way that these work on the SPI bus is that first you write the address of the register you want to read and then you read as many bytes as you need. The address is auto-incremented each time you read, so sending the first address followed by eight reads transfers the contents of all of the registers.

There is one other detail to note. The most significant bit of the address byte is used to indicate a read or write operation – it has to be set to one for a read. If it is zero then the bytes that you send following the address are stored in the registers – the data registers are read-only. Notice that all the registers already have bit 7 set in their usual addresses and there is no need to do anything extra for a read.

So a read operation is typically:

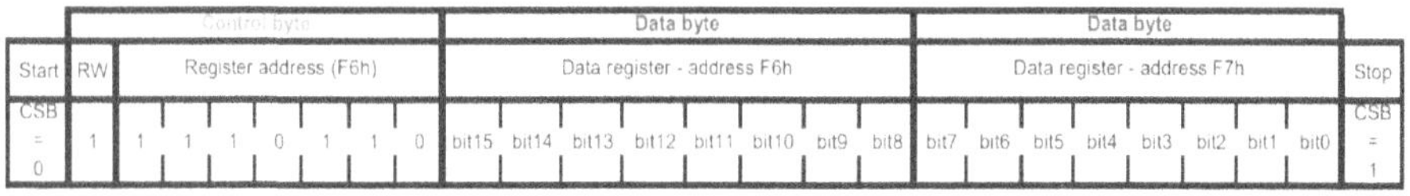

and a write operation is:

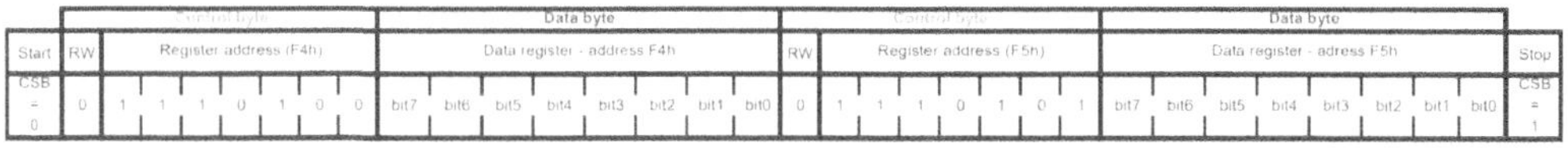

Notice that the operation is started by lowering the CS line and is stopped by raising the CS line. You can read or write as many values as you like and you can send another address to start the operation off at a different location.

If you look at the detailed specification of the SPI bus you will find that the CS line has to be low at least 20ns before the first clock pulse and has to remain low until at least 20ns after the final clock pulse. It also limits the clock to less than 10Mhz. It also uses mode zero by default.

This is enough information to start to create some functions that will read the BME280's raw data registers. First we need to set the SPI hardware and the GPIO lines correctly:

```
spi = SPI(0, sck=Pin(18), miso=Pin(16), mosi=Pin(19))
spi.init(baudrate=500_000, bits=8, polarity=0, phase=0,
                                          firstbit=SPI.MSB)
```

The CS line has to be set high before we start:

```
CS = Pin(17, Pin.OUT)
CS.high()
sleep_ms(1)
```

As a proof that the SPI bus is working, the first thing to try is to read the ID register which is at address `0xD0` and which returns a single byte giving the id number which should be `0x60` for the BME280 and `0x58` for the BMP280.

```
write = bytearray([0xD0])
CS.low()
spi.write(write)
read = spi.read(1)
CS.high()

print("Chip ID is",hex(read[0]))
spi.deinit()
```

As long as this works we can move on to read the raw data. If it doesn't work then either you have connected the device incorrectly or the device is broken.

We first have to set up what measurements we want to make and exactly how. This is controlled by two registers - `0xF4` controls pressure and temperature and `0xF2` controls humidity. You can see how the different bits in these registers set the accuracy and rate of measurement in the datasheet, but for general use we can set both registers using:

```
    CS.low()
write=bytearray([0xF2,0x01])
spi.write(write)
write=bytearray([0xF4,0x27])
spi.write(write)
CS.high()
```

This sets the humidity, temperature and pressure measurement to x1 over-sampling and normal mode sampling, which reads the sensors at a regular interval set by another register.

To read the raw data we can use the same technique of writing a register address, but this time we read eight items of data:

```
CS.low()
write=bytearray([0xF7])
spi.write(write)
sleep_ms(10)
rBuff = spi.read(8)
CS.high()
```

We can now put the bytes read from the device together to get the three readings:

```
pressure = (rBuff[0] << 12) |(rBuff[1] << 4) | (rBuff[2] >> 4)
temperature = (rBuff[3] << 12) | ( rBuff[4] << 4) | (rBuff[5] >> 4)
humidity = rBuff[6] << 8 | rBuff[7]
```

Putting all this together with some `print` statements to display the data gives the complete program:

```
from utime import sleep_ms
from machine import Pin, SPI
from time import sleep

spi = SPI(0, sck=Pin(18), miso=Pin(16), mosi=Pin(19))
spi.init(baudrate=500_000, bits=8, polarity=0, phase=0,
                                           firstbit=SPI.MSB)

CS = Pin(17, Pin.OUT)
CS.high()
sleep_ms(1)

write = bytearray([0xD0])
CS.low()
spi.write(write)
read = spi.read(1)
CS.high()
print("Chip ID is",hex(read[0]))

CS.low()
write=bytearray([0xF2,0x01])
spi.write(write)
write=bytearray([0xF4,0x27])
spi.write(write)
CS.high()

CS.low()
write=bytearray([0xF7])
spi.write(write)
sleep_ms(10)
rBuff = spi.read(8)
CS.high()

pressure = (rBuff[0] << 12) |(rBuff[1] << 4) | (rBuff[2] >> 4)
temperature = (rBuff[3] << 12) | ( rBuff[4] << 4) | (rBuff[5] >> 4)
humidity = rBuff[6] << 8 | rBuff[7]

print("Humidity = ", humidity)
print("Pressure = ", pressure)
print("Temp. = ", temperature)
```

If you try this out you will be disappointed to discover that the raw values aren't easy to interpret as real values. They are readings from the internal A-to-D converter and they have to be converted using the compensation formulas before they can be used.

Of course, it makes sense to factor out into functions the repetitive parts of the code to make it easier to extend the program. In particular, a `read_registers` function is sensible:

```
def read_registers(spi, CS, reg,n):
    CS.low()
    spi.write(bytearray([reg]))
    buf=spi.read(n)
    CS.high()
    return buf
```

If you want to make use of the BME280 for accurate measurements then it is essential that you read the compensation data and use it to compute corrected values. This isn't difficult, but the calculations are complicated and you can find them fully implemented in the example included in the C SDK examples. Now that you know the general logic of working with the device, the rest should be easier to understand as it is just a matter of implementing calculations.

Problems

The SPI bus is often a real headache because of the lack of a definitive standard, but in most cases you can make it work. The first problem is in discovering the characteristics of the slave device you want to work with. In general, this is solved by a careful reading of the datasheet or perhaps some trial and error, see the next chapter for an example.

If you are working with a single slave then generally things work once you have the SPI bus configuration set correctly. Things are more difficult when there are multiple devices on the same bus. The Pico has enough CS lines to handle eight devices, more if you only want to use a single SPI interface. Typically you will find SPI devices that don't switch off properly when they are not being addressed. In principle, all SPI devices should present high impedance outputs (i.e. tri-state buffers) when not being addressed, but some don't. If you encounter a problem you need to check that the selected slave is able to control the MISO line properly.

Summary

- The SPI bus is often problematic because there is no SPI standard.
- Unlike other serial buses, it makes use of unidirectional connections.
- The data lines are MOSI (master output slave input) and MISO (master input slave output).
- In addition, there is a clock line, output from master, and a number of select lines that you have to drive under program control.
- Timing for the select lines is a problem as you have to include a delay that makes sure that it remains low until the end of the final clock pulse.
- Data is transferred from the master to the slave and from the slave to the master on each clock pulse, arranged as a circular buffer.
- The Pico has two SPI devices which can work with almost any of the GPIO lines.
- The MicroPython provides all the functions you need to set up the SPI bus and transfer data one byte or multiple bytes at a time.
- You can test the SPI bus using a simple loopback connection.
- Working with a single slave is usually fairly easy, working with multiple slaves can be more of a problem.
- The BME280 is included as a complete example as part of the C SDK, but serves as a good example of how the SPI bus works.

Chapter 11

A-To-D and The SPI Bus

The SPI bus can be difficult to make work at first, but once you know what to look for about how the slave claims to work it gets easier. To demonstrate how it is done, let's add eight channels of 12-bit A-to-D using the MCP3008.

The Pico has a single A-to-D Converter (ADC) with analog inputs, four of which can be used via external pins. Before moving on to look at the use of SPI to add more A-to-D converters, let's take a look at those the Pico already has. If you are only interested in the detail of using the SPI bus, skip to that section.

Pico ADC

The Pico has a single onboard ADC which has four multiplexed inputs. Three are available on the external pins and one is used for an internal temperature sensor.

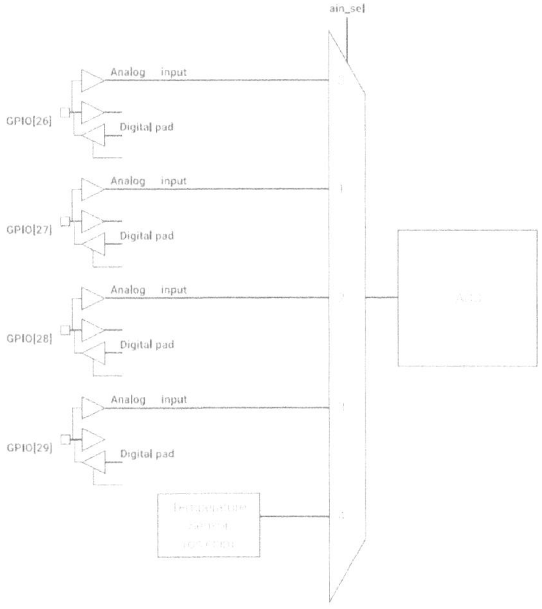

You can see from the diagram that pins GP26, GP27, GP28 and GP29 can be used as analog inputs. The ADC is a successive approximation converter. You don't need to know how it works to use it, but it isn't difficult to understand. The input voltage is compared to a standard voltage, VREF. The idea is that first a voltage equal to VREF/2 is generated and the input voltage is compared to this. If it is lower then the most significant bit is a 0 and if it is equal or greater then it is a 1. At the next step the voltage generated is VREF/2+VREF/4 and the comparison is repeated to generate the next bit. Successive approximation converters are easy to build, but they are slow. The Pico's ADC needs 96 clock cycles to create a 12-bit result, which gives a maximum sampling rate of 500kS/s with a default clock of 48MHz.

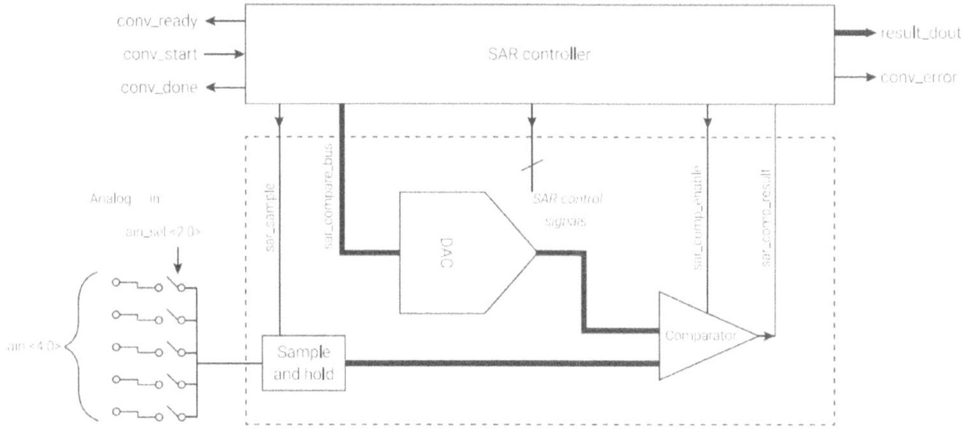

The Pico's ADC is very useful, but it isn't as accurate as you might hope. Although it has 12 bits, the claimed effective accuracy is only 9 bits and most of this loss of precision is due to the way the reference voltage is supplied and to noise due to its integration with other digital devices. According to the documentation, you can improve the accuracy by modifying the reference voltage:

> For much improved ADC performance, an external 3.0V shunt reference, such as LM4040, can be connected from the ADC_VREF pin to ground. Note that if doing this the ADC range is limited to 0-3.0V signals (rather than 0-3.3V), and the shunt reference will draw continuous current through the 200R filter resistor (3.3V-3.0V)/200 = ~1.5mA.

The ADC can be set to do conversions continuously, reading each of the possible inputs in a round-robin fashion, storing the results in an 8-deep FIFO (the documentation says 4-deep) buffer. Alternatively, you can simply start a conversion on a given input when you need the data – this is the only mode that MicroPython supports.

The simplest way of using the ADC is to perform a single read of a single input under software control. Before making use of it, you have to create an ADC object associated with the channel:

```
adc=machine.ADC(channel)
```

Once you have selected the input you are going to read, you can start a conversion and get the result using:

```
ADC.read_u16()
```

The result is in the range 0-65535 and this has to be scaled to give a physically meaningful reading.

The simplest A-to-D program you can write is:

```
import machine
import utime
sensor_temp = machine.ADC(4)
conversion_factor = 3.3 / (65535)

while True:
    reading = sensor_temp.read_u16() * conversion_factor
    temperature = 27 - (reading - 0.706)/0.001721
    print(temperature)
    utime.sleep(2)
```

This reads the temperature sensor which has a temperature voltage relationship given by:

$T = 27 - (V - 0.796)/0.001721$

You can create individual ADC objects for each channel and read them in nay order.

If you want to use any of the more advanced features of the Pico's ADC you will need to either create the functions that directly access the hardware or move to C.

The MCP3008 SPI ADC

An alternative to using the Pico's built-in ADC with its noise and stability problems is to use an external chip. The MCP3000 family is a low-cost versatile SPI-based set of A-to-D converters. Although the MCP3008, with eight analog inputs at 10-bit precision, and the MCP3004, with four analog inputs at 10-bit precision, are the best known, there are other devices in the family, including ones with 12-bit and 13-bit precision and differential inputs, at around the same sort of cost - $1 to $2.

In this chapter the MCP3008 is used because it is readily available and provides a good performance at low cost, but the other devices in the family work in the same way and could be easily substituted.

The MCP3008 is available in a number of different packages but the standard 16-pin PDIP is the easiest to work with using a prototyping board. You can buy it from the usual sources including Amazon, see Resources on this book's webpage. Its pinouts are fairly self-explanatory:

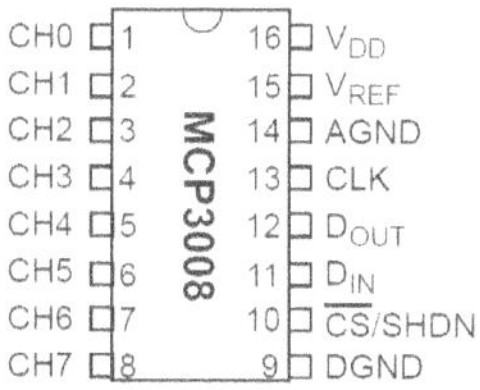

You can see that the analog inputs are on the left and the power and SPI bus connections are on the right. The conversion accuracy is claimed as 10 bits, but how many of these bits correspond to reality and how many are noise depends on how you design the layout of the circuit.

You need to take great care if you need high accuracy. For example, you will notice that there are two voltage inputs, VDD and VREF. VDD is the supply voltage that runs the chip and VREF is the reference voltage that is used to compare the input voltage. Obviously, if you want highest accuracy, VREF, which has to be lower than or equal to VDD, should be set by an accurate low-noise voltage source. However, in most applications VREF and VDD are simply connected together and the usual, low- quality, supply voltage is used as the reference. If this isn't good enough then you can use anything from a Zener diode to a precision voltage reference chip such as the TL431. At the very least, however, you should add a 1μF capacitor to ground connected to the VDD pin and the VREF pin.

The MC3000 family is based on the same type of ADC as the Pico's built-in device, a successive approximation converter.

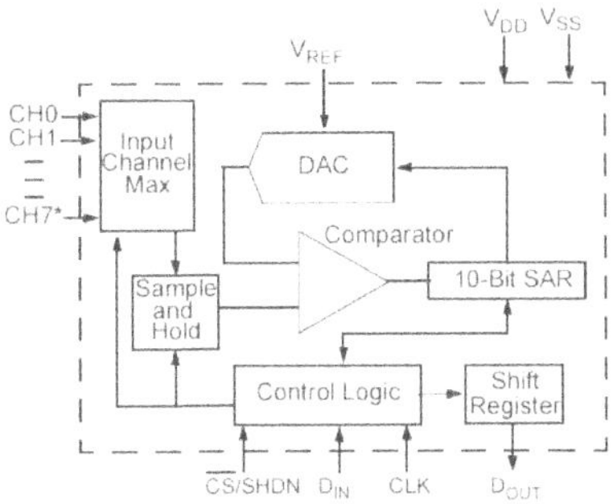

You can see that successive approximation fits in well with a serial bus as each bit can be obtained in the time needed to transmit the previous bit. However, the conversion is relatively slow and a sample-and-hold circuit has to be used to keep the input to the converter stage fixed. The sample-and-hold takes the form of a 20pF capacitor and a switch. The only reason you need to know about this is that the conversion has to be completed in a time that is short compared to the discharge time of the capacitor. So, for accuracy, there is a minimum SPI clock rate as well as a maximum.

Also, to charge the capacitor quickly enough for it to follow a changing voltage, it needs to be connected to a low-impedance source. In most cases this isn't a problem, but if it is you need to include an op amp. If you are using an op amp buffer then you might as well implement an anti-aliasing filter to remove frequencies from the signal that are too fast for the ADC to respond to. How all this works takes us into the realm of analog electronics and signal processing and well beyond the core subject matter of this book.

You can also use the A-to-D channels in pairs, i.e. in differential mode, to measure the voltage difference between them. For example, in differential mode you measure the difference between CH0 and CH1, i.e. what you measure is CH1-CH0. In most cases, you want to use all eight channels in single-ended mode. In principle, you can take 200k samples per second, but only at the upper limit of the supply voltage, i.e. VDD=5V, falling to 75k samples per second at its lower limit of VDD=2.7V.

The SPI clock limits are a maximum of 3.6MHz at 5V and 1.35MHz at 2.7V. The clock can go slower, but because of the problem with the sample-and-hold mentioned earlier, it shouldn't go below 10kHz. How fast we can take samples is discussed later in this chapter.

Connecting To The Pico

The connection from the MCP3008 to the Pico's SPI bus is very simple and can be seen in the diagram below.

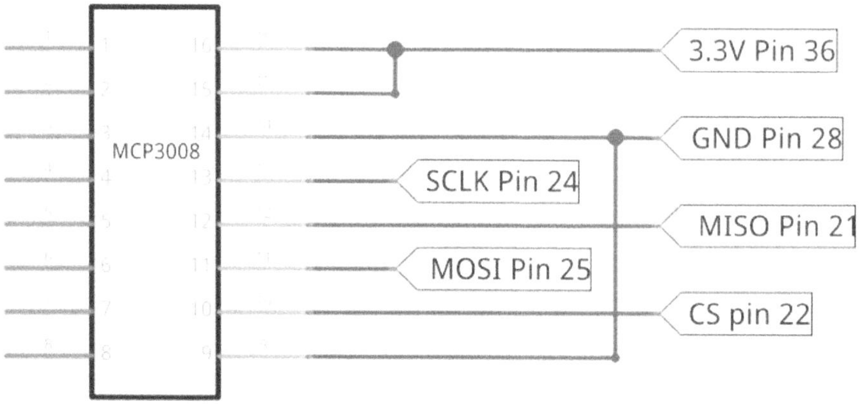

Pico	MCP3008
GP16 (pin 21) MISO	Pin 12
GP17 (pin 22) Chip Select	Pin 10
GP18 (pin 24) SCLK	Pin 13
GP19 (pin 25) MOSI	Pin 11
3.3v (pin 36)	Pins 15 and 16
GND (pin 38)	Pins 14 and 9

The only additional component that is recommended is a 1μF capacitor connected between pins 15 and 16 to ground, which is mounted as close to the chip as possible. As discussed in the previous section, you might want a separate voltage reference for pin 15 rather than just using the 3.3V supply.

Basic Configuration

Now we come to the configuration of the SPI bus. We have some rough figures for the SPI clock speed - around 10kHz to a little more than 1.35MHz. So a clock frequency of 500kHz seems a reasonable starting point.

From the datasheet, the chip select has to be active low and, by default, data is sent most significant bit first for both the master and the slave. The only puzzle is what mode to use? This is listed in the datasheet as mode `0 0` with clock active high or mode `1 1` with clock active low. For simplicity we will use mode `0 0`.

We now have enough information to initialize the slave:

```
spi = SPI(0, sck=Pin(18), miso=Pin(16), mosi=Pin(19))
spi.init(baudrate=500_000, bits=8, polarity=0, phase=0,
firstbit=SPI.MSB)

CS = Pin(17, Pin.OUT)
CS.high()
sleep_ms(1)
```

The Protocol

Now we have the SPI initialized and ready to transfer data, but what data do we transfer? As already discussed in the previous chapter, the SPI bus doesn't have any standard commands or addressing structure. Each device responds to data sent in different ways and sends data back in different ways. You simply have to read the datasheet to find out what the commands and responses are.

Reading the datasheet might be initially confusing because it says that you have to send five bits to the slave - a start bit, a bit that selects its operating mode single or differential, and a 3-bit channel number. The operating mode is 1 for single-ended and 0 for differential.

So to read Channel `3`, i.e. `011`, in single-ended mode you would send the slave:

```
11011xxx
```

where an `x` can take either value. In response, the slave holds its output in a high impedance state until the sixth clock pulse, then sends a zero bit on the seventh, followed by bit 9 of the data on the eighth clock pulse.

That is, the slave sends back:

```
xxxxxx0b9
```

where x means indeterminate.

The remaining nine bits are sent back in response to the next nine clock pulses. This means you have to transfer three bytes to get all ten bits of data. This all makes reading the data in 8-bit chunks confusing.

The datasheet suggests a different way of doing the job that delivers the data more neatly packed into three bytes. What it suggests to send a single byte is:

```
00000001
```

At the same time, the slave transfers random data, which is ignored. The final 1 is treated as the start bit. If you now transfer a second byte with most significant bit indicating single or differential mode, then a 3-bit channel address and the remaining bits set to 0, the slave will respond with the null and the top two bits of the conversion. Now all you have to do to get the final eight bits of data is to read a third byte:

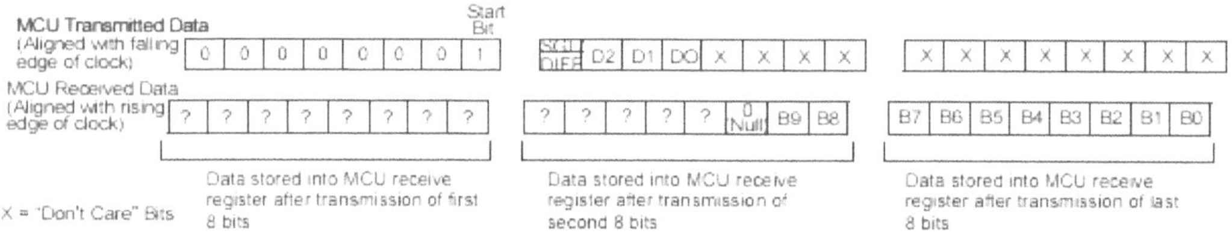

This way you get two neat bytes containing the data with all the low-order bits in their correct positions.

Using this information we can now write some instructions that read a given channel. For example, to read Channel 0 we first send a byte set to `0x01` as the start bit and ignore the byte the slave transfers. Next we send `0x80` to select single-ended and Channel 0 and keep the byte the slave sends back as the two high-order bits. Finally, we send a zero byte (`0x00`) so that we get the low-order bits from the slave:

```
CS.low()
write=bytearray([0x01, 0x80, 0x00])
read=bytearray(3)
spi.write_readinto(write,read)
CS.high()
```

Notice you cannot send the three bytes one at a time using `transfer` because that results in the CS line being deactivated between the transfer of each byte.

To get the data out of `rBuff` we need to do some bit manipulation:

```
data =  (read[1] & 0x03) << 8 |  read[2]
```

The first part of the expression extracts the low three bits from the first byte the slave sent and, as these are the most significant bits, they are shifted up eight places. The rest of the bits are then ORed with them to give the full 10-bit result. To convert to volts we use:

```
volts =  data * 3.3 / 1023.0
```

assuming that VREF is 3.3V.

In a real application you would also need to convert the voltage to some other quantity, like temperature or light level.

If you connect a logic analyzer to the SPI bus you will see something like:

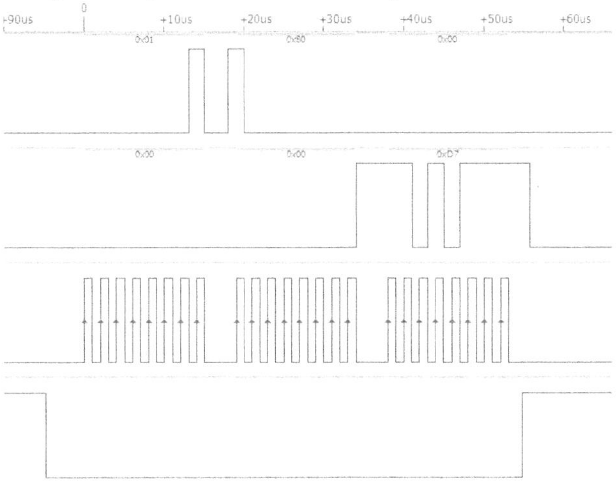

You can see the commands and the response, in this case a reading of 0.693V.

The complete program is:

```
from utime import sleep_ms
from machine import Pin, SPI
from time import sleep

spi = SPI(0, sck=Pin(18), miso=Pin(16), mosi=Pin(19))
spi.init(baudrate=500_000, bits=8, polarity=0, phase=0,
firstbit=SPI.MSB)
CS = Pin(17, Pin.OUT)
CS.high()
sleep_ms(1)
CS.low()
write=bytearray([0x01, 0x80, 0x00])
read=bytearray(3)
spi.write_readinto(write,read)
CS.high()
data =  (read[1] & 0x03) << 8 |  read[2]
volts =  data * 3.3 / 1023.0
print(volts)
spi.deinit()
```

SPI ADC Class

This all works, but it would be good to have a class with a method that read the ADC on a specified channel:

```
class spiADC:
    def __init__(self,spi,sckNo,misoNo,mosiNo,CSNo):
        self.spi = SPI(spi, sck=Pin(sckNo),
                    miso=Pin(misoNo), mosi=Pin(mosiNo))
        self.spi.init(baudrate=500_000, bits=8,
                    polarity=0, phase=0, firstbit=SPI.MSB)
        self.CS = Pin(CSNo, Pin.OUT)
        self.CS.high()
        sleep_ms(1)

    def read(self,chan):
        write=bytearray([0x01, (0x08 | chan) << 4 , 0x00])
        self.CS.low()
        read=bytearray(3)
        self.spi.write_readinto(write,read)
        self.CS.high()
        data =  (read[1] & 0x03) << 8 |  read[2]
        volts =  data * 3.3 / 1023.0
        return volts
```

With this class the main program is very simple:

```
adc=spiADC(0,18,16,19,17)
volts=adc.read(1)
print(volts)
```

How Fast?

Once you have the basic facilities working, the next question is always how fast does something work. In this case we need to know what sort of data rates we can achieve using this ADC. The simplest way of finding this out is to use the fastest read loop for a channel:

```
adc=spiADC(0,18,16,19,17)
while True:
    volts=adc.read(1)
```

With the clock of 500kHz the sampling rate is measured to be 4.5kHz which compares to the 17kHz achievable with a basic C program .

Increasing the clock rate to 1MHz pushes the sampling rate to 6kHz.

Also notice that as the clock rate goes up, you have to ensure that the voltage source is increasingly low-impedance to allow the sample-and-hold to charge in a short time.

Summary

- The Pico has a single ADC with four inputs. It is subject to a lot of noise from the power supply and circuits around it which reduces its accuracy.
- You can read individual input lines using the ADC class.
- The ADC class doesn't support many of the more advanced features of the Pico's ADC hardware.
- Making SPI work with any particular device has four steps:
  1. Discover how to connect the device to the SPI pins. This is a matter of identifying pinouts and mostly what chip selects are supported.
  2. Find out how to configure the Pi's SPI bus to work with the device. This is mostly a matter of clock speed and mode.
  3. Identify the commands that you need to send to the device to get it to do something and what data it sends back as a response.
  4. Find, or work out, the relationship between the raw reading, the voltage and the quantity the voltage represents.
- The MCP3000 range of A-to-D converters is very easy to use via SPI.
- You can read data at rates as fast as 6kHz.

Chapter 12
Using The I2C Bus

The I2C, standing for I-Squared-C or Inter IC, bus is one of the most useful ways of connecting moderately sophisticated sensors and peripherals to any processor. The only problem is that it can seem like a nightmarish confusion of hardware, low-level interaction and high-level software. There are few general introductions to the subject because at first sight every I2C device is different, but there are shared principles that can help you work out how to connect and talk to a new device.

The I2C bus is a serial bus that can be used to connect multiple devices to a controller. It is a simple bus that uses two active wires: one for data and one for a clock. Despite there being lots of problems in using the I2C bus, because it isn't well standardized and devices can conflict and generally do things in their own way, it is still commonly used and too useful to ignore.

The big problem in getting started with the I2C bus is that you will find it described at many different levels of detail, from the physical bus characteristics and protocol to the details of individual devices. It can be difficult to relate all of this together and produce a working project. In fact, you only need to know the general workings of the I2C bus, some general features of the protocol, and know the addresses and commands used by any particular device.

To explain and illustrate these idea we really do have to work with a particular device to make things concrete. However, the basic stages of getting things to work, the steps, the testing and verification, are more or less the same irrespective of the device.

I2C Hardware Basics

The I2C bus is very simple from the hardware point of view. It has just two signal lines, SDA and SCL, the data and clock lines respectively. Each of these lines is pulled up by a suitable resistor to the supply line at whatever voltage the devices are working - 3.3V and 5V are common choices. The size of the pull-up resistors isn't critical, but 4.7K is typical as shown in the circuit diagram.

You simply connect the SDA and SCL pins of each of the devices to the pull-up resistors. Of course, if any of the devices have built-in pull-up resistors you can omit the external resistors. More of a problem is if multiple devices each have pull-ups. In this case you need to disable all but one.

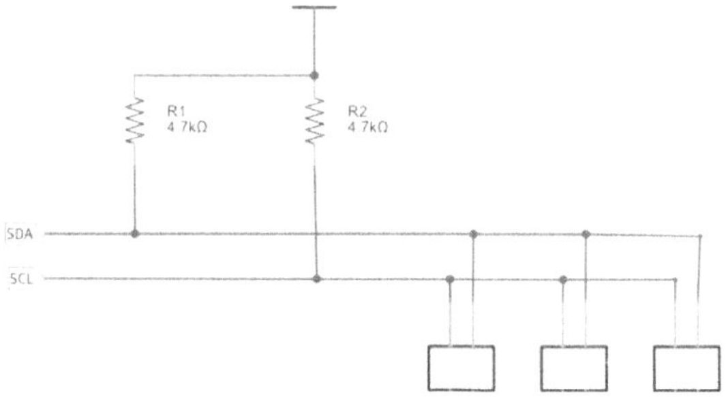

The I2C bus is an open collector bus. This means that it is actively pulled down by a transistor set to on. When the transistor is off, however, the bus returns to the high voltage state via the pull-up resistor. The advantage of this approach is that multiple devices can pull the bus low at the same time. That is, an open collector bus is low when one or more devices pulls it low and high when none of the devices is active.

The SCL line provides a clock which is used to set the speed of data transfer, one data bit is presented on the SDA line for each pulse on the SCL line. In all cases, the master drives the clock line to control how fast bits are transferred. The slave can, however, hold the clock line low if it needs to slow down the data transfer. In most cases the I2C bus has a single master device, the Pico in our case, which drives the clock and invites the slaves to receive or transmit data. Multiple masters are possible, but this is advanced and usually not necessary.

All you really need to know is that all communication usually occurs in 8-bit packets. The master sends a packet, an address frame, which contains the address of the slave it wants to interact with. Every slave has to have a unique address, which is usually 7 bits, but it can be 10 bits, and the Pico does support this. In the rest of this chapter we will use 7-bit addressing because it is commonly supported.

One of the problems in using the I2C bus is that manufacturers often use the same address, or same set of selectable addresses, and this can make using particular combinations of devices on the same bus difficult or impossible.

The 7-bit address is set as the high-order 7 bits in the byte and this can be confusing as an address that is stated as `0x40` in the datasheet results in `0x80` being sent to the device. The low-order bit of the address signals a write or a read operation depending on whether it is a `0` or a `1` respectively. After sending an address frame it then sends or receives data frames back from the slave. There are also special signals used to mark the start and end of an exchange of packets, but the library functions take care of these.

This is really all you need to know about I2C in general to get started, but it is worth finding out more of the details as you need them. You almost certainly will need them as you debug I2C programs.

The clock SCL and data SDA lines rest high. The master signals a Start bit by pulling the SDA line down – S in the diagram below. The clock is then pulled low by the master, during which time the SDA line can change state. The bit is read in the middle of the following high period of the clock pulse B1, B2 and so on in the diagram. This continues until the last bit has been sent when the data line is allowed to rise while the clock is high, so sending a stoP bit – P in the diagram. Notice that when data is being transmitted the data line doesn't change while the clock is high. Any change in the data line when the clock is high sends a start or a stop bit, i.e. clock high and falling data line is a start bit and clock high and rising data line is a stop bit:

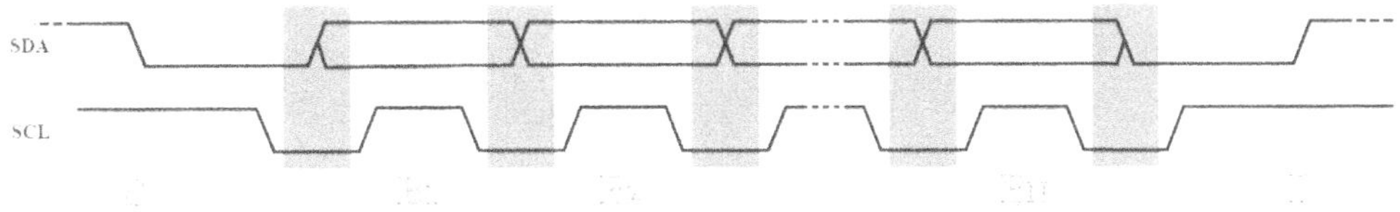

The clock speed was originally set at 100kHz, standard mode, but then increased to 400kHz in fast mode. In practice, devices usually specify a maximum clock speed that they will work with.

The Pico I2C

The Pico has two I2C controllers, I2C0 and I2C1, that can work as a master or a slave. The connections from both controllers can be routed to different pins:

I2C0	SDA	GP0	GP4	GP8	GP12	GP16	GP20		GP28
	SCL	GP1	GP5	GP9	GP13	GP17	GP21		

I2C1	SDA	GP2	GP6	GP10	GP14	GP18	GP22	GP26
	SCL	GP3	GP7	GP11	GP15	GP19		GP27

To use one of the controllers you have to select a pair of GPIO lines to act as SDA and SCL when you create an I2C object.

The I2C Functions

There are I2C functions for initialization, configuration and for writing and reading to the registers. Let's look at each group in turn.

Initialization

The constructor lets you set up a I2C object ready to use:

```
machine.I2C(id, scl=Pin, sda=Pin, freq=400000)
```

The `id` gives the I2C controller to use and the `scl` and `sda` parameters specify the `Pin` object to use for the I2C lines.

You can use the `init` method to set things up after the constructor, but at the time of writing it doesn't seem to work.

The Pico doesn't support the most recent high speed modes and the maximum clock rate is 1MHz. MicroPython doesn't support I2C in master mode, even if the Pico does.

There is also a `scan` method that will return a list of addresses that correspond to active I2C devices on the bus. Not all I2C devices respond well to a scan and the method's use is best avoided if possible.

To stop using I2C hardware use the `deinit` method.

Write

There are a two similar write methods. The most basic is:

```
I2C.writeto(addr, buf, stop=True)
```

The first parameter is the address of the device and `buf` is a byte array containing the data you want to send.

The final parameter, `stop`, needs some explanation. When you use this method it first sends an address frame, a byte containing the address of the device you specified. Notice that the 7-bit address has to be shifted into the topmost bits and the first bit has to be zeroed for a write operation. So when you write to a device with an address of `0x40`, you will see `0x80` on a logic analyzer, i.e. `0x40<<1`. After the address frame as many data frames are sent as specified in `src` and `len`.

The final parameters control how the stop bit is sent or not. The usual write transaction is:

```
START|ADDR|ACK|DATA0|ACK|
               DATA1|ACK|
                    ....
               DATAn|ACK|STOP
```

Notice that it is the slave that sends the `ACK` bit and, if the data is not received correctly, it can send `NAK` instead. Also notice that there is a single `STOP` bit at the end of the transaction and this is what you get if you set `stop` to True. If you set `stop` to False then the final stop bit isn't sent and the next data transfer can continue as part of the same transaction.

Notice that multibyte transfer is quite different from sending single bytes one at a time:

```
START| ADDR |ACK|DATA0|ACK|STOP
START| ADDR |ACK|DATA1|ACK|STOP
     ...
START| ADDR |ACK|DATAn|ACK|STOP
```

Notice that there are now multiple `ADDR` frames sent as well as multiple `START` and `STOP` bits. What this means in practice is that you have to look at a device's datasheet and send however many bytes it needs as a single operation. You cannot rely on being able to send the same number of bytes broken into chunks.

The other write function works in the same way, but with a slight variation.

```
I2C.writevto(addr, vector, stop=True)
```

This writes `vector`, which can be a List or a tuple of objects which support the Buffer Protocol. You can create custom objects for this or you can use `bytes` or a `bytearray`.

Writing To A Register

A very standard interaction between master and slave is writing data to a register. This isn't anything special and, as far as the I2C bus is concerned, you are simply writing raw data. However, datasheets and users tend to think in terms of reading and writing internal storage locations, i.e. registers in the device. In fact, many devices have lots of internal storage, indeed some I2C devices, for example I2C EPROMS, are nothing but internal storage.

In this case a standard transaction to write to a register is:

1. Send address frame
2. Send a data frame with the command to select the register
3. Send a data frame containing the byte, or word, to be written to the register

So, for example, you might use:

```
buf=bytearray([registerAddress,data])
i2c.writeto(addr,buf,stop= True)
```

Notice the command that has to be sent depends on the device and you have to look it up in its datasheet. Also notice that there is a single `START` and `STOP` bit at the beginning and end of the transaction.

To make working with memory devices easier, MicroPython provides some special methods:

```
I2C.writeto_mem(addr, memaddr, buf,  addrsize=8)
```

This writes to a device at `addr` and then access its memory location at `memaddr` writing the contents of `buf`. You don't need these methods as you can do the same job using more basic methods.

Read

The read functions are similar to the write functions. The most important is:

```
I2C.readfrom_into(addr, buf, stop=True)
```

and the parameters mean the same things as for the corresponding write function. This sends an address frame and then reads as many bytes from the slave as specified by the size of the `buf` bytearray. As in the case of `write`, the address supplied is shifted up one bit and the lower-order bit set to `1` to indicate a read operation. So, if the current slave is at address `0x40`, the read sends a read address of `0x81` – this is important to remember if you are viewing the transaction on a logic analyzer.

A simple alternative read method is:

```
buf=I2C.readfrom(addr, nbytes, stop=True)
```

which reads `nbytes` into a byte array.

The read transaction is:

```
START|ADDR|ACK|DATA0|ACK|
              |DATA1|ACK|
              |DATA2|ACK|
            ...
              |DATAn|NAK|STOP
```

The master sends the address frame and the slave sends the ACK after the address to acknowledge that it has been received and it is ready to send data. Then, the slave sends bytes, one at a time, and the master sends ACK in response to each byte. Finally, the master sends a NAK to indicate that the last byte has been read and then a STOP bit. That is, the master controls how many bytes are transferred.

As in the case of the write functions, a block transfer of n bytes is different from transferring n bytes one at a time and you can suppress the final stop bit by setting `nostop` to true.

Reading A Register

As for writing to a register, reading from a register is a very standard operation, but it is slightly more complicated in that you need both a write and a read operation. That is, to read a register you need a write operation to send the address of the register to the device and then a read operation to get the data that the device sends as the contents of the register.

So, for example, you would use something like:

```
buf=bytearray([registerAddress])
i2c.writeto(addr,buf,stop= True)
buf=i2c.readfrom(addr,1,stop=True)
```

If the register sends multiple bytes then you can usually read these one after another as a block transfer without sending an address frame each time. Notice that we don't suppress the stop bit between the read and the write to make it a single transaction.

In theory, and mostly in practice, a register read of this sort can work with a stop-start separating the write and the read operation, which is what you get if you use separate write and read function calls without suppressing the stop bit. That is, the transfer sequence is:

```
START|ADDR|ACK|REGADDR|ACK|STOP|
START|ADDR|ACK|DATA1|ACK|
              |DATA2|ACK|
               ...
              |DATAn|NAK|STOP
```

If you look at the end of the write and the start of the read using a logic analyzer, you will see that there is a STOP and START bit between them.

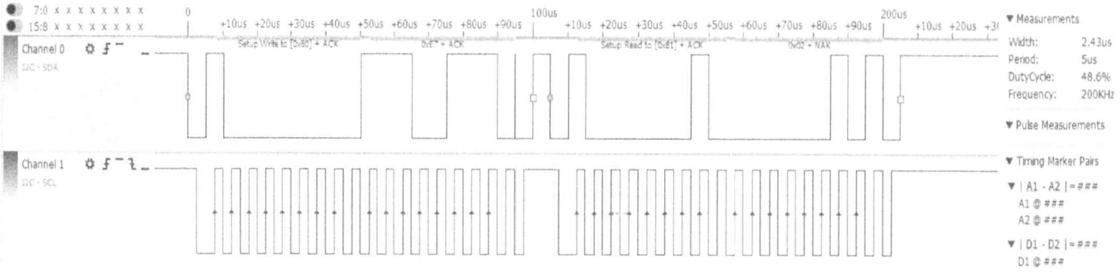

For some devices this is a problem. A STOP bit is a signal that another transaction can start and this might allow another master to take over the bus. To avoid this some devices demand a repeated START bit between the write and the read and no STOP bit. This is referred to as a "repeated start bit" or a "restart" transaction. The sequence for a repeated start bit register read is:

```
START|ADDR|ACK|REGADDR|ACK|
START|ADDR|ACK|DATA0|ACK|
              |DATA1|ACK|
       ...
              |DATAn|NAK|STOP
```

Notice that there is only one STOP bit.

In theory, either form of transaction should work, but in practice you will find that some slave devices state that they need a repeated start bit and no stop bits in continued transactions. In this case you need to be careful how you send and receive data. For example, to read a register from a device that requires repeated START bits but no STOP bit you would use:

```
buf=bytearray([registerAddress])
i2c.writeto(addr,buf,stop= False)
buf=i2c.readfrom(addr,1,stop=True)
```

You can see in the logic analyzer display that there is now just a single START bit between the write and the read.

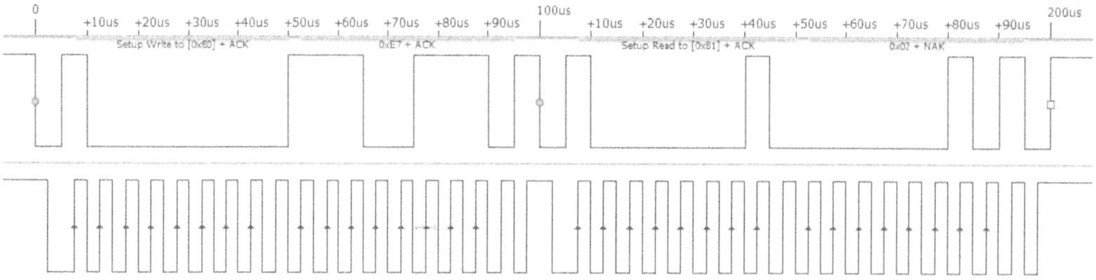

Very few devices need a repeated start transaction. The documentation mentions the MLX90620 IR array, but this is hardly a common peripheral. In practice, it usually doesn't make any difference if you send a stop bit in the middle of a write/read transaction, but you need to know about it just in case.

To make reading from a register easier there are two special methods:

```
I2C.readfrom_mem(addr, memaddr, nbytes, addrsize=8)
I2C.readfrom_mem_into(addr, memaddr, buf,  addrsize=8)
```

Both first write the `memaddr` to the device at `addr` and then perform a read of that device. They are equivalent to the write/reads given earlier.

Slow Read Protocols

The I2C clock is mostly controlled by the master and this raises the question of how we cope with the speed that a slave can or cannot respond to a request for data.

There are two broad approaches to waiting for data on the I2C bus. The first is simply to request the data and then perform reads in a polling loop. If the device isn't ready with the data, then it sends a data frame with a `NAK` bit set. In this case the read function throws an exception rather than return the number of bytes read. So all we have to do is test for an error response with a Try. Of course, the polling loop doesn't have to be "tight". The response time is often long enough to do other things and you can use the I2C bus to work with other slave devices while the one you activated gets on with trying to get the data you requested. All you have to do is to remember to read its data at some later time.

The second way is to allow the slave to hold the clock line low after the master has released it – so called “clock stretching”. In most cases the master will simply wait before moving on to the next frame while the clock line is held low. This is very simple and it means you don't have to implement a polling loop, but also notice that your program is frozen until the slave releases the clock line.

Many devices implement both types of slow read protocol and you can use whichever suits your application.

A Real Device

Using an I2C device has two problems - the physical connection between master and slave and figuring out what the software has to do to make it work. Here we'll work with the HTU21D/Si7021 and the information in its datasheet to make a working temperature humidity sensor using the I2C functions we've just met.

First the hardware. The HTU21D Humidity and Temperature sensor is one of the easiest of I2C devices to use. Its only problem is that it is only available as a surface-mount package. To overcome this you could solder some wires onto the pads or buy a general breakout board. However, it is much simpler to buy the HTU21D breakout board because this has easy connections and built-in pull-up resistors. The HTU21D has been replaced by the Si7021, which is more robust than the original and works in the same way, but the HTU21D is still available from many sources.

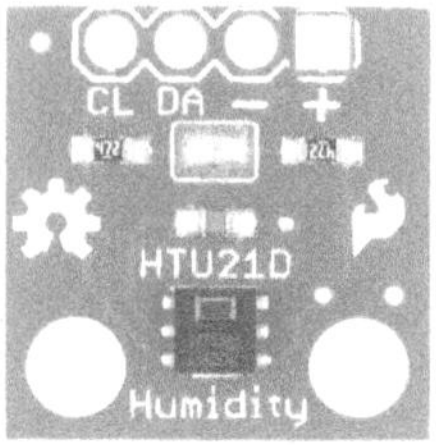

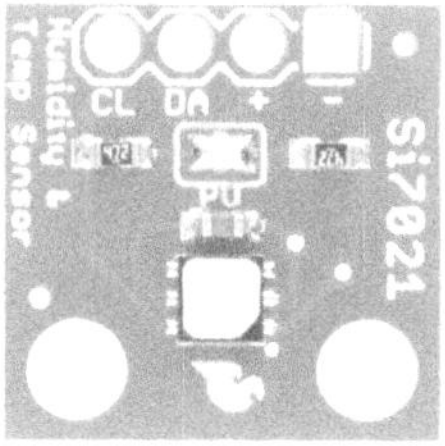

If you decide to work with some other I2C device you can still follow the steps given, modifying what you do to suit it. In particular, if you select a device that only works at 5V you might need a level converter.

Given that the HTU21D has pull-up resistors you don't need to enable the onboard pull-ups provided by the Pico. If you notice any irregularity in the signal at higher frequencies then adding some additional pull-ups might help.

You can use a prototype board to make the connections and this makes it easier to connect other instruments such as a logic analyzer. Given that the pinouts vary according to the exact make of the device, you need to compare the suggested wiring with the breakout board you are actually using.

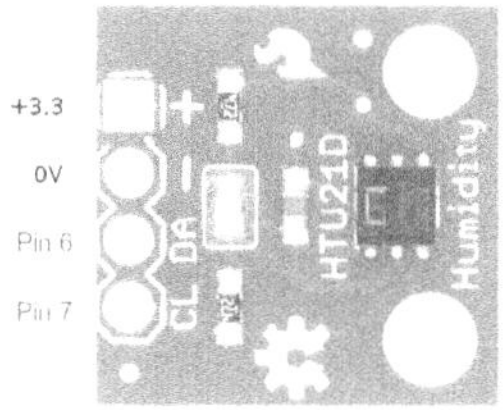

Pico	HTU21
SDA GP4 pin 6	SDA/DA
SCK GP5 pin 7	SCL/CL
3.3v pin 36	VCC/VIN/+
GND pin 38	GND/-

A First Program

After wiring up any I2C device, the first question that needs to be answered is, does it work? Unfortunately for most complex devices finding out if it works is a multi-step process. Our first program aims to read some data back from the HTU21D, any data will do.

If you look at the datasheet you will find that the device address is `0x40` and that it supports the following commands/registers:

Command	Code	Comment
Trigger Temperature Measurement	0xE3	Hold master
Trigger Humidity Measurement	0xE5	Hold master
Trigger Temperature Measurement	0xF3	No Hold master
Trigger Humidity Measurement	0xF5	No Hold master
Write user register	0xE6	
Read user register	0xE7	
Soft Reset	0xFE	

The easiest of these to get started with is the Read user register command. The user register gives the current setup of the device and can be used to set the resolution of the measurement.

Notice that the codes that you send to the device can be considered as addresses or commands. In this case you can think of sending `0xE7` as a command to read the register or the read address of the register, it makes no difference. In most cases, the term "command" is used when sending the code makes the device do something, and the term "address" is used when it simply makes the device read or write specific data.

To read the user register we have to write a byte containing `0xE7` and then read the byte the device sends back. This involves sending an address frame, a data frame, and then another address frame and reading a data frame. The device seems to be happy if you send a stop bit between each transaction or just a new start bit.

A program to read the user register is fairly easy to put together. The address of the device is `0x40`, so its write address is `0x80` and its read address is `0x81`. Recall that bus addresses are shifted one bit to the left and the base address is the write address and the read address is base address+1. As the I2C functions adjust the address as needed, we simply use `0x40` as the device's address, but it does affect what you see if you sample the data being exchanged:

```
from machine import Pin,I2C

i2c0=I2C(0,scl=Pin(17),sda=Pin(16),freq=400000)

buf = bytearray([0xE7])
i2c0.writeto( 0x40, buf, True)
read= i2c0.readfrom(0x40, 1, True)
print("User Register =",read)
```

This sends the address frame `0x80` and then the data byte `0xE7` to select the user register. Next it sends an address frame `0x81` to read the data.

If you run the program you will see:

```
User Register = 2
```

This is the default value of the register and it corresponds to a resolution of 12 bits and 14 bits for the humidity and temperature respectively and a supply voltage greater than 2.25V.

You can use the special read from a register method to do the same job:

```
read=i2c0.readfrom_mem(0x40,0xE7,1)
```

This does exactly the same as the previous code – it writes `0xE7` to device `0x40` and then reads one byte.

The I2C Protocol In Action

If you have a logic analyzer that can interpret the I2C protocol connected, what you will see is:

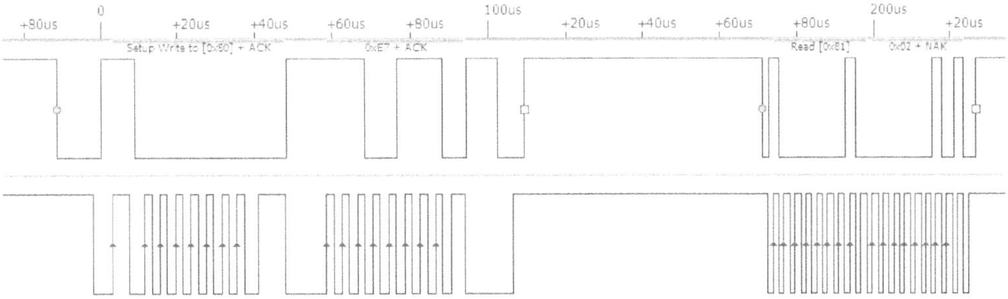

You can see that the `write_byte` function sends an address packet set to the device's 7-bit address `0x40` as the high-order bits with the low-order bit set to zero to indicate a write, i.e `0x80`. After this you get a data packet sent containing `0xE7`, the address of the register. After a few microseconds it sends the address frame again, only this time with the low-order bit set to one to indicate a read. It then receives back a single byte of data from the device, `0x02`. Also notice the start and stop bits at the end of each byte. The big gap between the write and the read is due to the time it takes MicroPython to process the method call. It is a limiting factor on how fast I2C can work.

This all demonstrates that the external device is working properly and we can move on to getting some data of interest.

Reading The Raw Temperature Data

Now we come to reading one of the two quantities that the device measures, temperature. If you look back at the command table you will see that there are two possible commands for reading the temperature:

Command	Code	Comment
Trigger Temperature Measurement	`0xE3`	Hold master
Trigger Temperature Measurement	`0xF3`	No Hold master

What is the difference between Hold master and No Hold master? This was discussed earlier in a general context under the section Slow Read Protocols. The device cannot read the temperature instantaneously and the master can either opt to be held waiting for the data, i.e. Hold master, or released to do something else and poll for the data until it is ready, i.e No Hold master.

The Hold master option works by allowing the device to stretch the clock pulse by holding the line low after the master has released it. In this mode the master will wait until the device releases the line. Not all masters support this mode, but the Pico does and this makes it the simpler option. To read the temperature using the Hold master mode you simply send `0xE3` and then read three bytes. The simplest program that will work is:

```
from utime import sleep_ms
from machine import Pin,I2C
from time import sleep

i2c0=I2C(0,scl=Pin(17),sda=Pin(16),freq=100*1000)

buf = bytearray([0xE3])
i2c0.writeto( 0x40, buf, False)
read= i2c0.readfrom(0x40, 3, True)
msb = read[0]
lsb = read[1]
check = read[2]
print("msb lsb checksum =", msb, lsb, check)
```

The read `bytearray` is unpacked into three variables with more meaningful names:
`msb` - most significant byte, `lsb` - least significant byte, `check` - checksum

If you try this out you should find that it works and it prints something like:

```
msb lsb checksum = 110 194 29
```

with the temperature in the 20°C range.

You can also use the read register method to do the same job:

```
read=i2c0.readfrom_mem(0x40, 0xE3, 3)
```

The logic analyzer reveals what is happening. First we send the usual address frame and write the `0xE3`. Then, after a short pause, the read address frame is sent and the clock line is held low by the device (lower trace):
The clock line is held low by the device for over 42ms while it gets the data ready. It is released and the three data frames are sent:

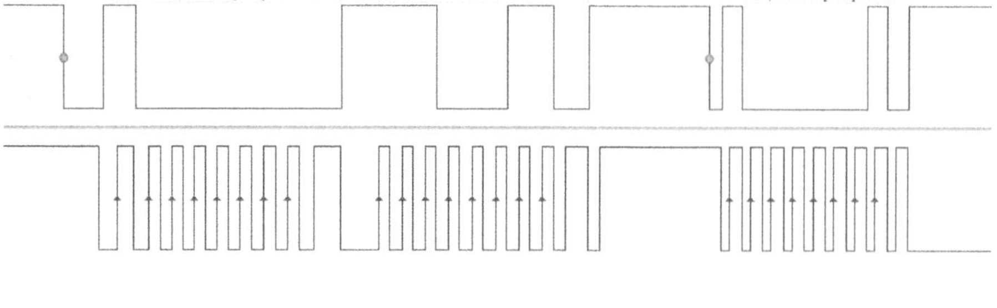

This response is a long way down the logic analyzer trace (40ms+) so keep scrolling until you find it.

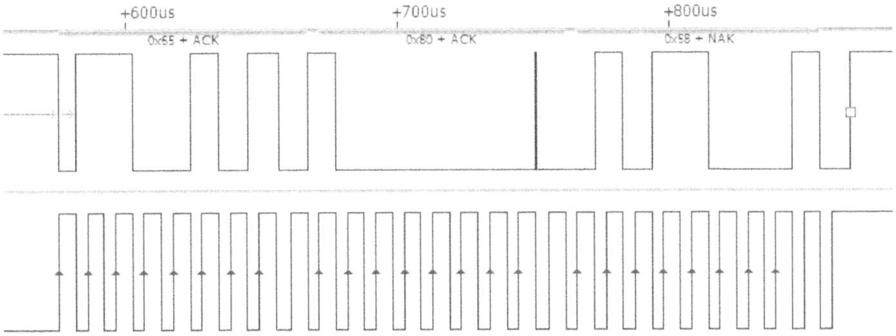

Notice that we suppress the stop bit between the write and the read to make it a single transaction.

Processing The Data

Our next task isn't really directly related to the problem of using the I2C bus, but it is a very typical next step. The device returns the data in three bytes, but the way that this data relates to the temperature isn't simple.

If you read the datasheet you will discover that the temperature data is the 14-bit value that results from putting together the most and least significant bytes and zeroing the bottom two bits. The bottom two bits are used as status bits, bit zero currently isn't used and bit one is a `1` if the data is a humidity measurement and a `0` if it is a temperature measurement.

To put the two bytes together we use:

```
data16= (msb << 8) |  (lsb & 0xFC)
```

This zeros the bottom two bits, shifts the `msb` up eight bits and ORs the two together. The result is a 16-bit temperature value with the bottom two bits zeroed. Now we have a raw temperature value but we have still have to convert it to standard units. The datasheet gives the formula:

Temperature in °C= -46.85 + 175.72 * data16 / 2^{16}

The only problem in implementing this is working out 2^{16}. You can work out 2^x with the expression `1<<x`, i.e. shift `1` `x` places to the right.

This gives:

```
temp = (-46.85 +(175.72 * data16 /(1<<16)))
```

Now all we have to do is print the temperature:

```
print("Temperature C ", temp)
```

The full listing is at the end of this chapter.

Reading Humidity

The nice thing about using I2C devices is that it gets easier. Once you have seen how to do it with one device, the skill generalizes and, once you know how to deal with a particular part of a device, other aspects of the device are usually similar.

While clock stretching is simple it sometimes doesn't work with some slave and master combinations. It is worth knowing how the alternative method works. For this reason let's implement the humidity reading using polling. We also find out how to use the No Hold master mode of reading the data, which is sometimes useful.

We write the 0xF5 once to the slave and then repeatedly attempt to read the three-byte response. If the slave isn't ready it simply replies with a NAK which the read method signals throwing an exception. We need to keep reading until the data is ready and this means using try to deal with the exception :

```
while True:
    sleep_ms(1)
    try:
        i2c0.readfrom_into(0x40,read, True)
        break
    except:
        continue
```

This polls repeatedly until the slave device returns an ACK, when the data is loaded into the buffer. Once we have the data, the formula to convert the 16-bit value to percentage humidity is:

```
RH= -6 + 125 * data16 / 216
```

Putting all this together, and reusing some variables from the previous parts of the program, we have:

```
buf = bytearray([0xF5])
i2c0.writeto( 0x40, buf, True)
read=bytearray(3)
while True:
    sleep_ms(1)
    try:
        i2c0.readfrom_into(0x40,read, True)
        break
    except:
        continue
msb = read[0]
lsb = read[1]
check = read[2]
print("msb lsb checksum =", msb, lsb, check)
data16 = (msb << 8) | (lsb & 0xFC)
hum = -6 + (125.0 * data16) / 65536
print("Humidity ", hum)
```

Checksum Calculation

Although computing a cyclic redundancy checksum, CRC, isn't specific to I2C, it is another common task. The datasheet explains that the polynomial used is:

$$X^8+X^5+X^4+1$$

Once you have this information you can work out the divisor by writing a binary number with a one in each location corresponding to a power of X in the polynomial. In this case the 8th, 5th, 4th and 1st bit. Hence the divisor is:

```
0x0131
```

What you do next is roughly the same for all CRCs. First you put the data that was used to compute the checksum together with the checksum value as the low-order bits:

```
data32 = (msb << 16)|(lsb <<8)| check
```

Now you have three bytes, i.e 24 bits, in a 32-bit variable. Next you adjust the divisor so that its most significant non-zero bit aligns with the most significant bit of the three bytes. As this divisor has a `1` at bit eight, it needs to be shifted 15 places to the right to move it to be the 24th bit:

```
divisor =  0x0131 <<15
```

or

```
divisor = 0x988000
```

Now that you have both the data and the divisor aligned, you step through the topmost 16 bits, i.e. you don't process the low-order eight bits which hold the received checksum. For each bit you check to see if it is a `1`. If it is you replace the data with the data XOR divisor. In either case you shift the divisor one place to the right:

```
for i in range(16):
       if data32 & 1<<(23 - i):
            data32 ^= divisor
       divisor>>= 1
```

When the loop ends, if there was no error, the `data32` should be zeroed and the received checksum is correct and as computed on the data received.

A complete function to compute the checksum, with some optimization, is:

```
def crcCheck(msb, lsb,check):
    data32 = (msb << 16)|(lsb <<8)| check
    divisor = 0x988000
    for i in range(16):
        if data32 & 1<<(23 - i):
            data32 ^= divisor
        divisor>>= 1
    return data32
```

It is rare to get a CRC error on an I2C bus unless it is overloaded or subject to a lot of noise.

Complete Listing

The complete program for reading temperature and humidity, including checksums, is:

```
from utime import sleep_ms
from machine import Pin,I2C
from time import sleep

def crcCheck(msb, lsb,check):
    data32 = (msb << 16)|(lsb <<8)| check
    divisor = 0x988000
    for i in range(16):
        if data32 & 1<<(23 - i):
            data32 ^= divisor
        divisor>>= 1
    return data32

i2c0=I2C(0,scl=Pin(17),sda=Pin(16),freq=100*1000)
buf = bytearray([0xE3])
i2c0.writeto( 0x40, buf, False)
read= i2c0.readfrom(0x40, 3, True)
msb = read[0]
lsb = read[1]
check = read[2]
print("msb lsb checksum =", msb, lsb, check)
data16= (msb << 8) |  (lsb & 0xFC)
temp = (-46.85 +(175.72 * data16 /(1<<16)))
print("Temperature C ", temp)
print("Checksum=",crcCheck(msb,lsb,check))

buf = bytearray([0xF5])
i2c0.writeto( 0x40, buf, True)
read=bytearray(3)
while True:
    sleep_ms(1)
    try:
        i2c0.readfrom_into(0x40,read, True)
        break
    except:
        continue
msb = read[0]
lsb = read[1]
check = read[2]
print("msb lsb checksum =", msb, lsb, check)
data16 = (msb << 8) | (lsb & 0xFC)
hum = -6 + (125.0 * data16) / 65536
print("Humidity ", hum)
print("Checksum=",crcCheck(msb,lsb,check))
```

Notice that we used clock stretching in reading the temperature and polling in reading the humidity. In practice, you would choose one method according to your needs.

Of course, this is just the start. Once you have the device working and supplying data, it is time to write your code in the form of functions that return the temperature and the humidity and generally make the whole thing more useful and easier to maintain. This is often how this sort of programming goes. First you write a lot of inline code so that it works as fast as it can, then you move blocks of code to functions to make the program more elegant and easy to maintain, checking at each refactoring that it all still works.

Not all devices used standard bus protocols. In Chapter 14 we'll look at a custom serial protocol that we have to implement for ourselves.

Summary

- The I2C bus is simple yet flexible and is one of the most commonly encountered ways of connecting devices.
- The I2C bus uses two wires – a data line and a clock.
- The Pico has two I2C interfaces.
- Each I2C interface can be connected to a pair of GPIO lines.
- The I2C protocol isn't standardized and you have to take account of variations in the way devices implement it.
- There are single byte transfer operations and multibyte transfers which differ in when a stop bit is sent.
- The low-level protocol can be made slightly more high-level by thinking of it as a single write or read a register operation.
- Sometimes a device cannot respond immediately and needs to keep the master waiting for data. There are two ways to do this, polling and clock stretching.
- The HTU21D is a simple I2C device, but getting it working involves using polling or stretching.
- Computing a checksum is an involved, but common, operation.

Chapter 13
Using The PIO

The most interesting and attractive feature of the Pico is arguably its PIO – Programmable I/O hardware. So far we have used direct access to the GPIO lines or to some built-in hardware that controls them such as PWM, I2C or SPI. If you want to connect to a device that these standards apply to then your problem is solved. If, on the other hand, you have a device that doesn't use one of these standards then you are faced with the task of constructing your own implementation of the interface. For example, the DHT22 temperature and humidity sensor uses its own protocol for delivering its data and the 1-Wire bus is commonly encountered, but not often supported, in hardware. You can even go beyond any standard and implement your own custom data transfer protocol.

The traditional way to approach any custom or unsupported protocol is to use "bit banging". This is simply the act of writing a program which controls GPIO lines to simulate the hardware that might be used to implement the protocol. Basically what you have to do is set GPIO lines high and low as dictated by the timing of the protocol and then read data at set times. This is easy in theory, but getting the timings right is harder that it appears and synchronization between state changes is particularly challenging. In addition there is the problem that bit banging ties up the processor. In many cases this doesn't matter. In other cases you can use one core to implement the protocol and another to process the data. However the Pico's PIOs provide a general-purpose solution to the problem that offers advantages that go well beyond the traditional solutions.

The Pico documentation on the PIO even suggests that if you find yourself working on a bit-banging solution then you should stop and start again using the PIO. I wouldn't go this far because the PIO solution is more complex and specialized and if a simple bit-banging solution works you need a motivation to transfer your attention to the PIO. This said, it is worth spending time learning about the PIO if you anticipate ever making use of it. With the right approach it isn't as difficult as it first appears.

PIO Basic Concepts

The problem with getting to grips with the PIO is that there are two distinct views of it – inside and outside. You can also add to this the extra complication of setting up a project that makes use of it, but that is a one-time problem. The best way to think about the PIO is as a black box that performs some transaction using GPIO lines and presents and accepts data from the processor – exactly like other I/O subsystems PWM, SPI, I2C etc. The only difference is that the transaction it performs is programmable.

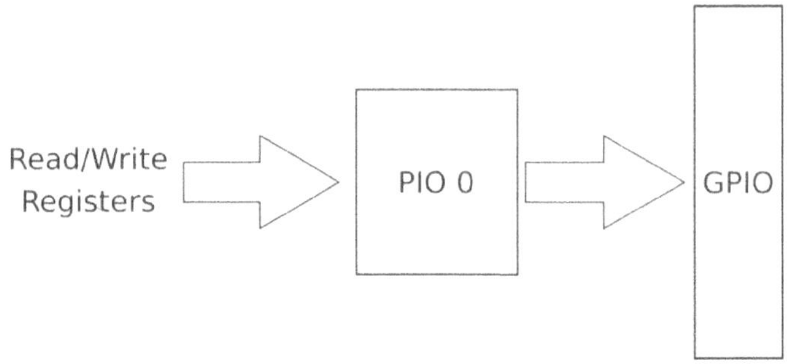

This the same situation as for the I2C or SPI hardware – once set up, you communicate with the PIO by reading and writing registers and it is associated with a number of GPIO lines.

As you might guess from the illustration there is more than one PIO. The Pico has two, PIO0 and PIO1, and each can be programmed independently of the other. To make things simple, we will use PIO0, but everything translates to PIO1 if you need to use it.

You communicate with the PIO by reading and writing registers and, as with other hardware modules, MicroPython provides functions and methods to do this for you.

There are two independent PIOs and each one has four state machines. The state machines can be set to run a program stored in the PIO. It is the state machines that deal with the outside world and the PIO deals with your program. Each PIO only has storage for 32 instructions, which the four state machines share. A common configuration is for each state machine to run the same program, but work with different GPIO lines. If the programs are small then it is also possible for each state machine to have its own program stored in the PIO.

Although MicroPython doesn't really distinguish between the state machines, it is worth keeping in mind that state machines 0 to 3 are part of PIO0 and state machines 4 to 7 are part of PIO1.

State Machines And GPIO

The first and most important thing to understand is that the each state machine doesn't have instructions that work with particular GPIO lines. You don't write a PIO program that sets the output of pin GP2, say. Instead you specify which GPIO lines are to be included in a number of groups. Each state machine has its own set of GPIO lines that it works with and, at the simplest, each state machine can perform the same sort of task on different GPIO pins. So, for example, if you programmed the PIO to be an SPI controller then you could define four different SPI interfaces, each to a different set of pins. It is also possible to run each state machine from a different program, but more of this later.

Each state machine has two straightforward sets of GPIO output lines, OUT and SET that are specified by a base and a count, e.g. base is GP4 and count is 2 gives GP4 and GP5. Lines in the OUT group are controlled by instructions that transfer data from the processor to the state machine. That is, they are lines you set by writing data to the state machine and then using instructions to transfer it to the outside world. Typically the lines in the OUT group are what you would think of as data lines.

Those in the SET group are controlled by instructions running on the state machine. That is, they are lines you set within the PIO program and typically are used as clock lines.

A third group of output lines, SIDESET, is also available. While it turns out to be very useful it is slightly more difficult to understand. We will return to it later.

The important point is that these groups of lines are not set in your program but as part of the initialization of the PIO. This means that your program can be reused with different GPIO lines on different state machines.

To work with the PIO there are two new elements in MicroPython, both in the rp2 module. The first is the `rp2.asm_pio` function which is intended to be used as a decorator applied to a function that defines the PIO program, i.e. the PIO assembly language. The second is the `StateMachine` class which configures the state machine you are going to use. The key idea is that the PIO program should be specified so that it can be run on any state machine with a specific set of GPIO lines. In other words, there are configuration options that are specific to the program, e.g. the number of GPIO lines used, and there are options which are specific to the state machine, e.g. exactly which lines are used.

The program-specific options are parameters within the `rp2.asm_pio` decorator. It is worth making a list of these here because currently they are mostly undocumented, even though they will be explained as we go along:

`set_init`	Set the initial states and hence number of the SET group
`sideset_init`	Set the initial states and hence number of the SETSIDE group
`out_init`	Set the initial states and hence number of the OUT group
`autopull`	Enable or disable autopull
`autopush`	Enable or disable autopush

The state machine-specific options are set in the `StateMachine` constructor:

`freq`	Clock frequency in Hz
`in_base`	Start GPIO pin of IN Group
`out_base`	Start GPIO pin of OUT Group
`set_base`	Start GPIO pin of SET Group
`sideset_base`	Start GPIO pin of SIDESET Group
`jmp_pin`	GPIO pin used for `jmp` instruction
`in_shiftdir`	ISR shift direction using `rp2.PIO.SHIFT_RIGHT/LEFT`
`out_shiftdir`	OSR shift direction using `rp2.PIO.SHIFT_RIGHT/LEFT`
`push_thresh`	Autopush threshold
`pull_thresh`	Autopull threshold

All of this will make sense by the end of the chapter.

PIO Blinky

Now it is time to write and run our first PIO program and, in the grand tradition, we will eventually, write a Blinky program – the Hello World of hardware.

The PIO is programmed using a special assembly language and we will return to its main features later, but for now you can most probably understand the meaning of our first simple program:

```
.program squarewave
again:
    set pins, 1      ; Drive pin high
    set pins, 0      ; Drive pin low
    jmp again        ; Set PC to label `again`
```

The first instruction sets pin 1 in the SET group of pins to be an active output, i.e. it changes from a high impedance state to an output. This is more complicated than it seems. More of exactly how this works in the section on output to the GPIO.

The loop that follows this repeatedly changes the SET group of pins from high to low. Notice that from this program you cannot tell which pins are actually being used – this is set externally to the PIO program.

If you were working with C then the next step would be to use the PIO assembler to convert the program to PIO machine code. With MicroPython, however, you can enter the assembly language as a function and use a decorator to convert it into machine code:

```
@rp2.asm_pio(set_init=rp2.PIO.OUT_LOW)
def blink():
    label("again")
    set(pins, 1)
    set(pins, 0)
    jmp("again")
```

You can see that each of the assembly language instructions is written as a function call and the parameters of the function are the parameters of the assembly language instruction. Notice that we have a `label` function to mark a location in the program. In most cases you can write your PIO assembler using the assembly language and then convert it into function calls, but after a little practice you should find it easy to work directly in function calls. This is the approach used in the rest of this book. Notice that the PIO program's name is the name of the MicroPython function.

The `set_init` parameter in the decorator defines the SET group to have a single GPIO line which is initially set low. In general this would be a tuple of pins and initial states. This is how you determine the number of lines in each of the groups and their initial states.

Now we have the code as a function, the next job is to pick one of the eight possible state machines to execute and configure it. As already mentioned, the first four state machines, 0 to 3, belong to PIO0 and the second four, 4 to 7 belong to PIO1. This is important because all of the state machines in a given PIO share a limited 32-instruction program memory and which PIO is used to host the program depends on the state machine you choose.

To setup a state machine to run the program we use:

```
sm = rp2.StateMachine(0, blink, freq=2000, set_base=Pin(16))
```

This selects state machine zero, which is hosted by PIO0. It loads the `blink` program into PIO0 and sets state machine `0` to start running it using a clock frequency of 2000Hz and the start of the SET group as GP16. This combined with the program decorator:

```
@rp2.asm_pio(set_init=rp2.PIO.OUT_LOW)
```

makes the SET group just the single GPIO line, i.e. GP16.

To set the state machine running the program all we have to do is:

```
sm.active(1)
```

and to stop it running:

```
sm.active(0)
```

If you now look at the output of GP16 you will see a square wave.

The complete program is:

```
import rp2
from machine import Pin

@rp2.asm_pio(set_init=rp2.PIO.OUT_LOW, )
def blink():
    label("again")
    set(pins, 1)
    set(pins, 0)
    jmp("again")


sm = rp2.StateMachine(0, blink, freq=2000, set_base=Pin(16))
sm.active(1)
```

We do have a square wave-producing program but it isn't suitable as a Blinky example. At the moment the clock rate is set to produce a square wave at around 666Hz and we need to bring down the frequency to something more reasonable so that we can see an LED blink on and off.

Clock Division And Timing

There are a number of ways of controlling the speed with which things happen, the most direct and preferable is to change the clock frequency by specifying a divider. The PIO clock can be divided down using a 16-bit divider – 8-bit integer/8-bit fractional. In most cases, you should avoid using a fractional divider as it introduces extra periods to make the frequency average out to the fractional part. This can make the PIO subject to jitter, which decreases the reliability.

Using MicroPython you don't set the clock divider but its frequency and the system attempts to work out a clock divider that gets as close to the desired frequency as possible. The lowest clock frequency using the default clock, is about 2kHz which is too fast for some things. This brings us to another important idea. Most of the timing in PIO programs is linked to the time an instruction takes to complete and most instructions take one clock cycle. This can cause difficulties in that execution paths may not be equal in time when you want them to be. For example, consider our simple PIO program:

```
.program squarewave
again:
    set pins, 1      ; Drive pin high
    set pins, 0      ; Drive pin low
    jmp again        ; Set PC to label `again`
```

The first set pins takes one clock cycle, the second takes one clock cycle, but the jmp instruction also takes one clock cycle, meaning that the GPIO line is high for one clock cycle and low for two clock cycles.

There are a number of ways of changing timings. For example, you could include a nop instruction which wastes a single clock cycle, but also uses one memory slot. Alternatively you could specify a delay as part of the instruction – all PIO instructions can have a delay in clock cycles specified as [n] at the end of the instruction. So our new program:

```
.program squarewave
again:
    set pins, 1 [1]  ; Drive pin high
    set pins, 0      ; Drive pin low
    jmp again        ; Set PC to label `again`
```

Holds the line high for two clock cycles and low for two clock cycles.

With the maximum clock divider this gives a square wave output at around 120kHz which is still too fast for flashing an LED. The maximum delay you can specify in any instruction is 31 clock cycles, so even:

```
.program squarewave
again:
 set pins, 1 [31]  ; Drive pin high
 set pins, 0 [30]  ; Drive pin low
 jmp again         ; Set PC to label `again`
```

or, in function form:

```
@rp2.asm_pio(set_init=rp2.PIO.OUT_LOW, )
def blink():
    label("again")
    set(pins, 1) [31]
    set(pins, 0) [30]
    jmp("again")
```

only brings the clock down to 36Hz.

In general, you should try to choose a clock frequency that makes timing of pulses using just per-instruction delays to adjust the relative timings. Each instruction and delay takes one clock cycle, so this is the natural unit of measurement to use.

In our case we need a much longer delay than can be produced by32 clock cycles – we need a spin wait loop.

Writing Loops

One of the things you will most likely have to do is write the equivalent of a `for` or `while` loop in PIO assembler and it might not be obvious how to do this because it doesn't have control structures like MicroPython. To write a loop you need to know that each state machine has two scratch 32-bit storage registers called `x` and `y`. You can store values and use values in these registers from various sources using the `in`, `out`, `set` or `mov` instructions. If you want to set a register, `x` say, to a constant value then you need to use:

```
set x,31
```

where 31 is the maximum value you can set because the instruction is limited to a 5-bit immediate value.

You can use the registers to implement a for loop with up to 32 repeats. The `jmp` instruction can test the value of a register for zero and auto-decrement it after the test. This means an n-repeat for loop can be constructed as:

```
set x,n
loop:
    nop
jump x--, loop
```

Using this you can slow the rate at which the GPIO line is toggled by repeating the `nop` 32 times. However, even if you use:

```
set x,n
loop:
    nop [31]
jump x--, loop
```

the effective rate of toggling the GPIO line is still too great to see an LED flash. To slow it down even more you need two nested loops in the PIO program:

```
.program squarewave
    again:
     set pins, 1
     set x,31
     loop1:
       nop [31]
      jmp x--,loop1
      pins, 0
      set(x,31)
     loop2:
       nop [31]
       jmp x--,loop2
    jmp again
```

You can see that there are two delay loops. If you try it you will find it runs at about 1.1Hz and if you change the GPIO line to GP25 you will see the onboard LED flash.

The complete program is:

```
import rp2
from machine import Pin

@rp2.asm_pio(set_init=rp2.PIO.OUT_LOW, )
def blink():
    label("again")
    set(pins, 1)
    set(x,31)
    label("loop1")
    nop() [31]
    jmp(x_dec,"loop1")
    set(pins, 0)
    set(x,31)
    label("loop2")
    nop() [31]
    jmp(x_dec,"loop2")
    jmp("again")

sm = rp2.StateMachine(0, blink, freq=2000, set_base=Pin(16))
sm.active(1)
```

Notice that `x--` is entered as `x_dec` and other conditionals are entered in similar ways.

Also notice that we can only set the `x` or `y` register to a maximum of 31 because the PIO assembly instructions only have space for a 5-bit literal field. There is another way to set the `x` and `y` register using data transferred into the PIO program from the MicroPython program and this brings us to the subject of data input.

Data To The PIO

Just like any of the I/O subsystems in the Pico, you can send data to the PIO from your MicroPython program. The Pico provides a 3-element, 32-bit FIFO buffer, the TX FIFO, for input to the state machine and, usually, output to the GPIO lines. You can write data to the TX FIFO using:

```
sm.put(data)
```

It stores data in the TX FIFO and return immediately, unless there is no space when it waits for the PIO to read some data and free some space.

The PIO program can read data from the TX FIFO using the `pull` instruction which moves the first item of data from the FIFO into the Output Shift Register, OSR. Usually the OSR is then used to drive the OUT group of lines, more of this later. However, the mov instruction can be used to move the OSR to a range of destinations including the x and y registers. This means we can use the TX FIFO buffer to set the x or y register to a full 32-bit value.

Using this we can create a Blinky program that just uses one loop and can create time delays in excess of 30 minutes. The PIO program is:

```
.program squarewave
    pull block
again:
    set pins, 1
    mov x, osr
 loop1:
    jmp x--,loop1
   set pins, 0
   mov x, osr
loop2:
 jmp x--,loop2
jmp again
```

Notice the way the OSR is used as a way of storing the loop count so that it can be used to reset the x register each time. If you needed to use the OSR for something else this wouldn't be possible and you would have to use the y register or some other method. Also notice that the pull block instruction causes the PIO program to wait until there is some data in the TX FIFO to start things off.

The complete program is:

```
import rp2
from machine import Pin

@rp2.asm_pio(set_init=rp2.PIO.OUT_LOW, )
def squarewave():
    pull(block)
    label("again")
    set(pins, 1)
    mov(x,osr)
    label("loop1")
    jmp(x_dec,"loop1")
    set(pins, 0)
    mov(x,osr)
    label("loop2")
    jmp(x_dec,"loop2")
    jmp("again")

sm = rp2.StateMachine(0, squarewave, freq=2000, set_base=Pin(16))
sm.active(1)
sm.put(0xFFF)
```

The only real difference is the `put` function at the end which stores `0xFFF` in the TX FIFO to start the PIO program running. If you run this program you will find that the frequency is 0.3Hz. If you change the pin from GP16 to GP25 you will also see the onboard LED flash.
This is a way of passing a single parameter to a PIO program. If you want to pass two or more parameters then it can be done, but things become more complicated.

Output To GPIO

The use of the Output Shift Register, OSR, to store a value that is moved to the `x` register is reasonable, but it isn't the usual role for the OSR, which is normally used as the link between values passed to the state machine and the state of the GPIO lines. We have already seen that the SET group of GPIO lines can be controlled using the `set` command, now we move on to consider how the OUT group of GPIO lines can be controlled using the `out` command. Recall that the OUT group is a set of GPIO lines specified by a starting number and number of lines.

The data in the OSR can be moved to the OUT group of pins using the `out` instruction:

```
out pins, n
```

where `n` is the number of bits shifted to the OUT group. The bits that are shifted out are used to set the first `n` GPIO lines in the OUT group and any remaining lines are set to zero. Notice that while we use the term "shift" this applies only to what happens to the contents of the OSR register. The `n` bits are presented to the `n` GPIO lines in one operation – i.e. all of the GPIO lines change their state at the same time. The "shift" simply means that the data in the OSR register is moved to remove the `n` bits used. You can set the direction of the shift when you create the state machine so as to make either the most significant or least significant bits the ones that are used in the OSR register first:

```
out_shiftdir= rp2.PIO.SHIFT_RIGHT/LEFT
```

In general, 32 bits are sent to the destination of the `out` command, the lower `n` bits coming from the OSR and the remaining bits being zero. Of course, usually only the lower `n` bits are actually useful. If you repeat the `out` instruction the next `n` bits are used and so on until the shift register is empty when all 32 bits have been shifted out.

A simple PIO program example will help clarify. In this case we will send the bits from a 32-bit value to two GPIO pins just to show how things work:

```
.program output
    pull block
again:
    out pins,2
    jmp again
```

You can see that all the program does is to pull a single 32-bit value from the TX FIFO into the OSR and then uses the `out` instruction to send pairs of bits to the first two pins in the OUT group of pins. Of course, we need to set the number and start GPIO pin of the OUT group. The complete program is:

```
import rp2
from machine import Pin

@rp2.asm_pio(out_init=(rp2.PIO.OUT_LOW,rp2.PIO.OUT_LOW))
def output():
    pull(block)
    label("again")
    out(pins,2)
    jmp("again")

sm = rp2.StateMachine(0, output, freq=2000, out_base=Pin(16),
out_shiftdir=rp2.PIO.SHIFT_RIGHT)
sm.active(1)
sm.put(0xFEDCBA98)
```

You can see that we set the size and initial state of the OUT group to two pins set to low in the decorator and the start pin GP16. These two settings make the OUT group consist of GP16 and GP17. The final instruction sends some arbitrary data to the TX FIFO and this is output two bits at a time to the two GPIO lines, GP16 and GP17:

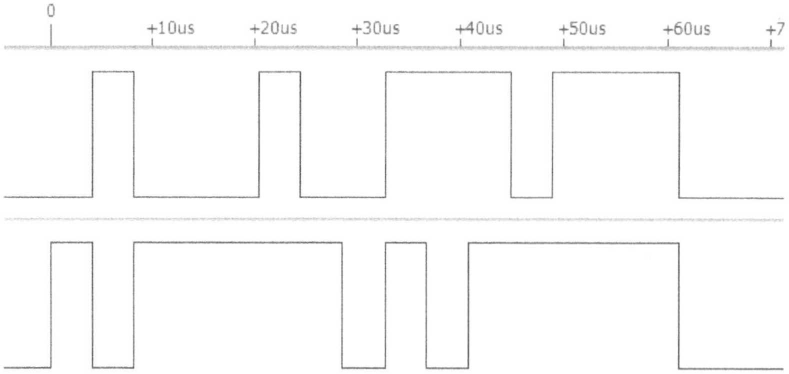

The output stops when the OSR is empty but the out instruction keeps sending zeros to the GPIO lines. Also notice the use of:

```
out_shiftdir=rp2.PIO.SHIFT_RIGHT
```

to set the direction of the shift. Try changing it to rp2.PIO.SHIFT_LEFT.

If you want to keep the data flowing, you have to keep refilling the OSR. For example, you could change the program to send data to the buffer whenever there was a free space:

```
while True:
    sm.put(0xFEDCBA98)
```

Now the PIO program can read new data from the TX FIFO whenever it wants to:

```
.program output
    pull block

again:
  out pins,2
  out pins,2
  out pins,2
  pull block
  jmp again
```

In this case we send the first three pairs of bits to the output pins and then read the TX FIFO into the OSR to get new data.

Of course, usually you would use all of the data in the OSR before reloading it. This standard task can be done automatically with the autopull which reloads the OSR whenever its contents fall below a threshold. You set autopull in the @rp2.asm_pio decorator:

```
autopull=True
```

If you set autopull to true then the OSR will be reloaded from the TX FIFO when it contents fall below the pull threshold which is set in the state machine constructor using:

```
pull_thresh=pt
```

Using this, the PIO program becomes:

```
.program output
again:
  out pins,2
  jmp again
```

The complete program is:

```
import rp2
from machine import Pin

@rp2.asm_pio(out_init=(rp2.PIO.OUT_LOW,rp2.PIO.OUT_LOW),autopull=Tru
e)
def output():
    label("again")
    out(pins,2)
    jmp("again")

sm = rp2.StateMachine(0, output, freq=2000, out_base=Pin(16),
out_shiftdir=rp2.PIO.SHIFT_RIGHT)
sm.active(1)
while True:
    sm.put(0xFEDCBA98)
```

There are some minor points of how `autopull` works we've not covered, but you now have the general principles. Notice that we are using the default pull threshold of 32 bits.

Side Effects

Now we have encountered most of the ideas behind PIO output to the GPIO lines, but there is one additional feature – lines in the SETSIDE group acting as a side effect. As well as being able to explicitly set GPIO lines using `set` or `out`, we can also arrange for any instruction to set GPIO lines as part of its execution. In this sense, the setting of the GPIO lines is a "side effect" of the instruction. Side effects are usually thought of as something to avoid in a standard program – you want an instruction to do what it appears to do and nothing more. In the case of PIO programming, however, side effects are very useful because they allow you to set GPIO lines according to where the program is in its execution. That is, they can be used to signal to the outside world the state of the program. At a more utilitarian level, being able to set a GPIO line as a side effect of another instruction can save both time and instructions.

The only downside of side effects is that they are coded using the space in the PIO instruction that is normally used to specify a delay. There is a 5-bit field used to code the delay and the side effect and you can trade bits for each purpose. The default is that all five bits are used to specify a delay and this makes it possible to specify up to 31 clock cycles of delay.

You can allocate bits to be used to SETSIDES using the PIO instruction:

```
.side_set count opt pindirs
```

where `count` is the number of bits to allocate, `opt` means that the setting is optional on an instruction and `pindirs` is the option to set the pin directions as a side effect. Setting `pindirs` as a side effect is useful if you need to change a line from `out` to `in` as the program progresses.

In MicroPython you configure the program using the decorator's parameter:

```
pio(sideset_init=rp2.PIO.OUT_LOW)
```

which gives the number of pins in the SETSIDE group and in the state machine using the:

```
setsideset_base=Pin(16)
```

parameter which sets the starting pin for the group.

Obviously you should choose to use as few bits as possible if you want to make use of a delay, but in MicroPython you can mostly ignore this and allow the system to determine it for you.

Finally to include a GPIO line to be set as a side effect you can add:

```
side value
```

after any PIO instruction and the value will be used to set the SETSIDE pins (or their directions).

To show how all this works the simplest thing to do is write the square wave generator again, but this time using nothing but side_set options:

```
.program squarewave
.side_set 1 opt
again:
  nop side 1
  jmp  again side 0
```

The `side_set` directive specifies a single bit, which means that the SETSIDE group of GPIO lines is effectively a single line. The `nop` has a side effect of setting the side effect line to `1` and the `jmp` sets it back to `0`.

When you translate this into MicroPython you need to know that the `side n` addition is converted into `.side(n)` and you determine the SETSIDE group using the state machine constructor and decorator.

The complete program for the squarewave is:

```
import rp2
from machine import Pin

@rp2.asm_pio(sideset_init=rp2.PIO.OUT_LOW)
def squarewave():
    label("again")
    nop().side(1)
    jmp("again").side(0)

sm = rp2.StateMachine(0, squarewave, freq=2000,sideset_base=Pin(16))
sm.active(1)
```

If you run this program you will find that pin GP16 is toggled in a 50% square wave at around 1kHz, without a single `out` or `set` instruction in sight. Notice that the rising edge occurs when the loop starts and the falling edge when the loop ends.

As an example of using more than one GPIO line as a side effect consider:

```
import rp2
from machine import Pin

@rp2.asm_pio(sideset_init=(rp2.PIO.OUT_LOW,rp2.PIO.OUT_LOW))
def squarewave():
    label("again")
    nop().side(2)
    jmp("again").side(1)

sm = rp2.StateMachine(0, squarewave, freq=2000,sideset_base=Pin(16))
sm.active(1)
```

Now we have set aside two bits to control SETSIDE GPIO lines, which means we can control two of them. First we write `2`, `10` in binary, to the pair of lines and then `1`, or `01`, which switches the lines in anti-phase – when one is high the other is low. Notice that the value written is a mask used to set the SETSIDE group of lines.

SETSIDE lines can overlap with OUT and SET lines and if such instructions set the same GPIO line, the sideset instruction has precedent.

Input

Now we have described most of the general ideas of PIO output to GPIO lines, it is time to consider input from GPIO lines and using the IN group. The good news is that this is easy to understand as long as you have followed the ideas involved in output. The only problem is that, in practice, input is always harder than output and, in the case of PIO programming, you have to re-think how things work. The reason is that you don't have any edge-triggered events or timers. Everything you do has to be conditioned on the state of the system and raw timings based on how long instructions take to execute.

The basic mechanisms of input are the same as output. There is a 4-element, 32-bit RX FIFO buffer that you can read from your MicroPython program using:

```
data=sm.get()
```

Obviously the function blocks until there is some data to read.

Data gets into the RX FIFO from the PIO via the Input Shift Register, ISR, which is the input equivalent of the OSR. To transfer data from the GPIO lines you use the `in` instruction,

```
in pins,n
```

The `in` instruction moves `n` bits from the IN GPIO group into the ISR shifting up any data that is already present. You can also use it to transfer `n` bits from the `x` or `y` registers and a few other sources. Notice that in MicroPython `in` has to become `in_` to avoid the clash with the Python keyword `in`. The IN group of GPIO lines is defined in the state machine constructor using:

```
in_base=Pin(n)
```

The IN group is treated differently from the rest. You only have to specify a base number as you can read from any GPIO line, even if it is an output. What this means is that an `in` instruction can read all of the 32 GPIO lines and only the numbering of the lines changes. For example, after:

```
sm = rp2.StateMachine(0, light, freq=2000,in_base=Pin(16))
```

which sets the base to GP16, the instruction:

```
in pins,3
```

will transfer the state of GP16, GP17 and GP18 to the ISR. The pin states are transferred irrespective of whether the pins are set to input or output. Another difference between the IN group and the others is that you have to initialize the GPIO lines used for input as a separate task as there is no `in_init` parameter.

Once you have sufficient bits in the ISR, you can transfer the entire 32-bit value to the RX FIFO using the `push` instruction:

```
push block
```

This waits for a free space in the RX FIFO and clears the ISR ready for further use. You can also use `autopush` to transfer data to the RX FIFO automatically when a bit count threshold is reached. You can set the direction in which the ISR is shifted and control the use of auto-push and its threshold using:

```
autopush=True
```

in the decorator and:

```
in_shiftdir
push_thresh=pt
```

in the state machine constructor.

As the simplest example, let's read GP2 and use its state to control the on-board LED. This isn't a useful program, but it is a good illustration of the basics. The PIO program is simply:

```
.program light
again:
  in pins,1
  push block
  jmp  again
```

You can see that this simply reads the state of the first INPUT pin and pushes it to the RX FIFO. Of course, after four reads and pushes, the RX FIFO will be full and the program will stall unless the FIFO is read by the main program.

The complete program simply has to set things up, read the data and set the on-board LED accordingly:

```
import rp2
from machine import Pin
@rp2.asm_pio()
def light():
    label("again")
    in_(pins,1)
    push(block)
    jmp("again")
LED=Pin(25,mode=Pin.OUT)
in1=Pin(16,mode=Pin.IN)
sm = rp2.StateMachine(0, light, freq=2000,in_base=Pin(16))
sm.active(1)
while True:
    flag = sm.get()
    if (flag == 0):
        LED.value(0)
    else:
        LED.value(1)
```

Notice the use of `in_base=Pin(16)` to set GP16 to be the start of the IN group, but it still has to be set to input using:

```
in1=Pin(16,mode=Pin.IN)
```

If you set GP16 to high and low you will see the LED turn on and off. The problem with the program is that you have no idea when it reads the input line. All you can say is that the speed of reading is controlled by how fast the program makes room for new data in the RX FIFO.

Edges

This is all there is to using input, but the big problem is when do you read the state of the GPIO lines? When you are implementing a data transfer protocol, it is usual that you output something and then wait a given time before reading. Alternatively, data is read in response to a clock state transition, often an edge, but we have no direct way of responding to an input edge.

The solution to the problem takes us a little way beyond the simple `in` instruction. The PIO supports a conditional wait and a conditional jump, based on the state of a GPIO line. The instruction:

```
wait state pin n
```

will wait until the pin indexed by `n` in the INPUT group is in the specified state. Notice that `n=0` is the first pin in the group.

You can also select the pin by absolute GPIO number:

```
wait state gpio n
```

This will wait until GP*n* is in the specified state. You can also wait on an IRQ, but this is beyond the scope of this first look at the PIO system.

The conditional jump:

```
jmp pin target
```

jumps to the target if the pin specified by *n* in the state machine constructor

```
jmp_pin=n
```

is a `1` and `pin` is a raw GPIO number and not specified by the INPUT group.

The `wait` instruction is about synchronizing the program to the outside world and the `jmp` is about determining what processing should occur according to the state of the outside world. For example, if you need to start a process based on the start of a rising edge you could use:

```
wait 0 pin 0
wait 1 pin 0
```

This works by waiting for the GPIO line to go to zero and hence, when the second wait ends, you know that a rising edge has just occurred. For this to be accurate the clock rate should be high compared to the pulses being input. Notice that if you read the line after the second wait, the state will be one clock cycle after the rising edge. In general, whatever you do as a result of the edge, will occur some number of clock cycles after the edge. This need for a high clock rate to localize the edge can make other aspects of timing difficult.

For example, we can generate a pulse that is close to the rising edge of an input pulse train. The PIO program is:

```
.program squarewave
again:
   wait 0 pin 0
   wait 1 pin 0
   set pins, 1
   set pins, 0
jmp  again
```

This waits on the first pin in the INPUT group to change from `0` to `1`, i.e. a rising edge, and then toggles the first pin in the SET group.

The complete program is:

```
import rp2
from machine import Pin

@rp2.asm_pio(set_init=rp2.PIO.OUT_LOW)
def squarewave():
    label("again")
    wait(0,pin, 0)
    wait(1,pin, 0)
    set(pins, 1)
    set(pins, 0)
    jmp("again")

in1=Pin(16,mode=Pin.IN)
sm = rp2.StateMachine(0, squarewave,
freq=2000,in_base=Pin(16),set_base=Pin(17))
sm.active(1)
```

Notice we have to set the `in_base` and `set_base` for the IN and SET groups.

With the clock set to a low rate, you can see that the marker pulse isn't very close to the rising edge:

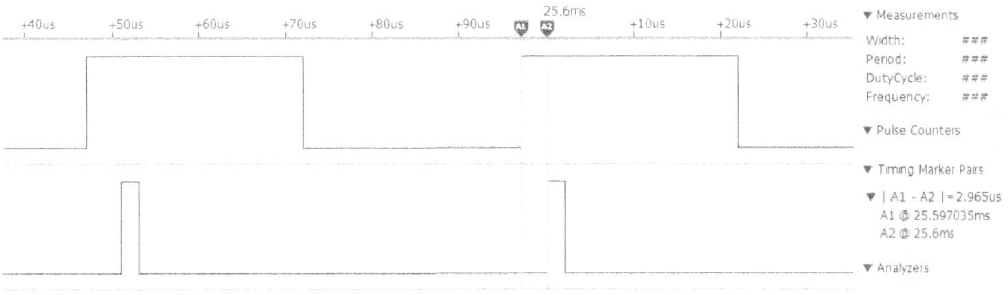

If you increase the clock rate to its maximum, the displacement becomes very small – 45n:

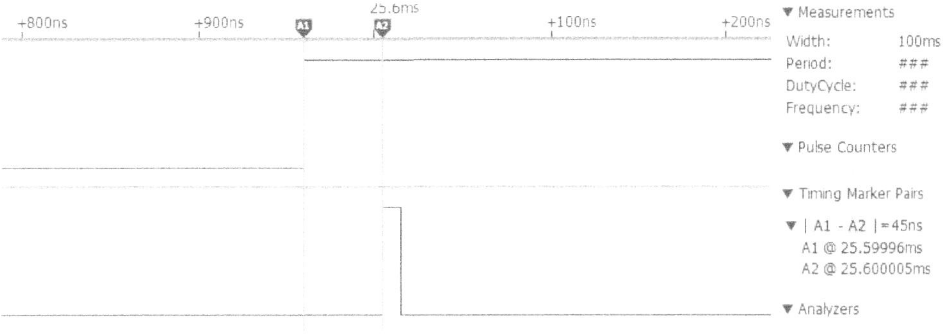

To see the problem, consider the task of moving the marker to the middle of the pulse where you might want to sample the line's state. At maximum clock rate this might be too many cycles to use a simple delay. A slower clock rate is easier, but less accurate in its alignment with the edge.

Advanced PIO

This first look at the PIO and its use has covered the main topics to make it possible for you to write PIO programs and use them in your projects. It isn't complete and there are many additional features that you still have to discover. Most of these features are about optimization. For example, nearly all PIO programs take the form of an infinite loop and this is a waste of one of the scarce 32 instruction locations and introduces a clock cycle delay.

As an alternative you can place:

```
.wrap_target
```

before the instruction that you want to restart the program at and:

```
.wrap
```

after the last instruction. For example, our previous program can be written

```
.program squarewave
.wrap_target
 wait 0 pin 0
 wait 1 pin 0
 set pins, 1
 set pins, 0
.wrap
```

The corresponding MicroPython function is:

```
@rp2.asm_pio(set_init=rp2.PIO.OUT_LOW)
def squarewave():
    wrap_target()
    wait(0,pin, 0)
    wait(1,pin, 0)
    set(pins, 1)
    set(pins, 0)
    wrap()
```

In this case there is no `jmp` instruction as the program counter is automatically set back to the target when it reaches the `wrap`. It is as if the program's address space was circular.

Other topics not discussed are the use of DMA, interrupts and the interaction between multiple PIOs. This is all very straightforward and logical, once you have mastered the basics.

Now we move on to discover how the PIO can do useful work for us.

Summary

- The PIO and the state machine are special processors designed to interact with the outside world.
- You can use a PIO attempt to implement any otherwise unsupported protocol.
- The Pico has two PIOs, each with four state machines.
- The GPIO lines associated with the PIO are determined by a set of groups – OUT, SET and SETSIDE for output and IN for input. GPIO lines also have to be set to PIO mode before they will work in any of the groups.
- You can set the clock frequency that the PIO uses to execute instructions one per clock cycle.
- The clock should be set to a frequency that is suitable for the sort of pulses the PIO is working with.
- It is easy to toggle a GPIO line, but slightly harder to make it slow enough to flash an LED. To do this you need to implement a spin wait loop.
- The OSR and ISR are used to send data to and receive data from the GPIO lines.
- There are two FIFO stacks which can be used to send data to the OSR and ISR from the processor.
- Every instruction can change the state of GPIO lines in the SETSIDE group as a side effect of its execution.
- Working with edges isn't natural for the state machine, but it can be achieved using wait instructions.

Chapter 14

The DHT22 Sensor Implementing A Custom Protocol

In this chapter we make use of all the ideas introduced in earlier chapters to create a raw interface with the low-cost DHT11/22 temperature and humidity sensor. It is an exercise in implementing a custom protocol directly in MicroPython using bit banging and then using the PIO. Given the documentation advises against using it, you might be wondering why we start with bit banging? The answer is that bit banging is easier to debug and it is usually a good idea to implement it first if possible, if only to prove that the device in question works and you know how it works.

The DHT22

The DHT22 is a more accurate version of the DHT11 and it is used in this project. The software will work with both versions and also with the AM2302, which is equivalent to the DHT22.

Model AM2302/DHT22
Power supply 3.3-5.5V DC
Output signal digital signal via 1-wire bus
Sensing element Polymer humidity capacitor
Operating range
 humidity 0-100%RH;
 temperature -40~80Celsius
Accuracy
 humidity +-2%RH(Max +-5%RH);
 temperature +-0.5Celsius
Resolution or sensitivity
 humidity 0.1%RH;
 temperature 0.1Celsius
Repeatability
 humidity +-1%RH;
 temperature +-0.2Celsius

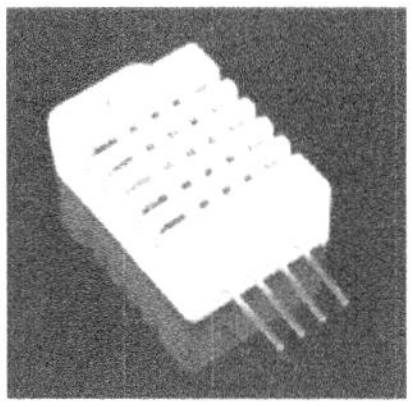

The device will work at 3.3V and it makes use of a 1-wire open collector-style bus, which makes it very easy to make the physical connection to the Pico.

The 1-wire bus used isn't standard and is only used by this family of devices, so we have little choice but to implement the protocol in MicroPython.

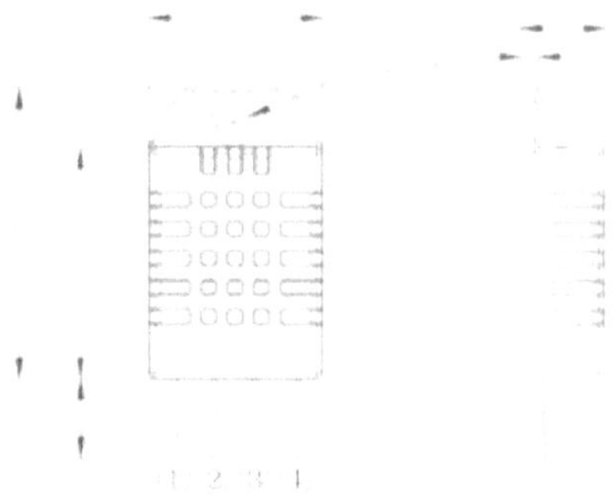

The pinouts are:

1. VDD
2. SDA serial data
3. Not used
4. GND

and the standard way of connecting the device is:

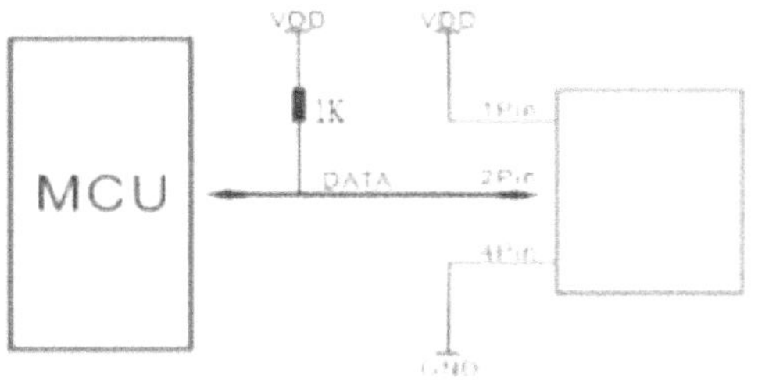

Although the recommended pull-up resistor is 1K, a higher value works better with the Pico - typically 4.7K, but larger will work.

The serial protocol is also fairly simple:

1. The host pulls the line low for between 0.8ms and 29ms, usually 1ms.
2. It then releases the bus which is pulled high.
3. After between 20μs and 200μs, usually 30μs, the device starts to send data by pulling the line down for around 80μs and then lets it float high for another 80μs.
4. Next 40 bits of data are sent using a 70μs high for a `1` and a 26μs high for a `0` with the high pulses separated by around 50μs low periods.

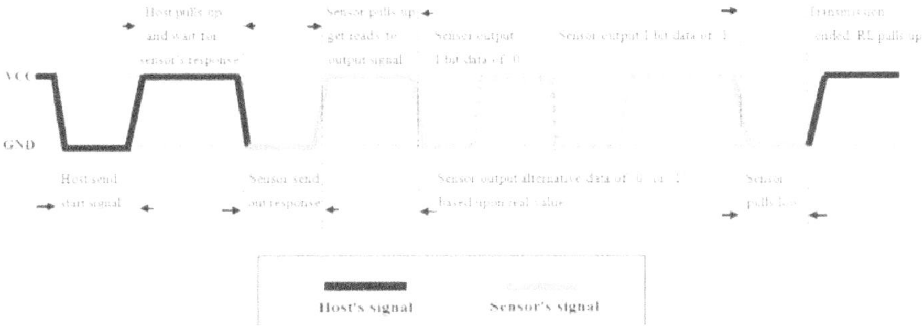

What we have to do is pull the line low for 1ms or so to start the device sending data and this is very easy. Then we have to wait for the device to pull the line down and let it pull up again for about 160μs and then read the time that the line is high or low 40 times.

A `1` corresponds to 70μs and a `0` corresponds to 26 to 28μs. This is within the range of pulse measurements that can be achieved using standard library functions. There is also a 50μs low period between each data bit and this can be used to do some limited processing. The time between falling edge transitions is therefore 120μs for a `1` and 76μs for a `0`.

When trying to work out how to decode a new protocol it often helps to try to answer the question, "how can I tell the difference between a `0` and a `1`?"

If you have a logic analyzer it can help to look at the waveform and see how you work it out manually. In this case, despite the complex-looking timing diagram, the difference comes down to a short versus a long pulse!

The Electronics

Exactly how you build the circuit is a matter of preference. The basic layout can be seen below.

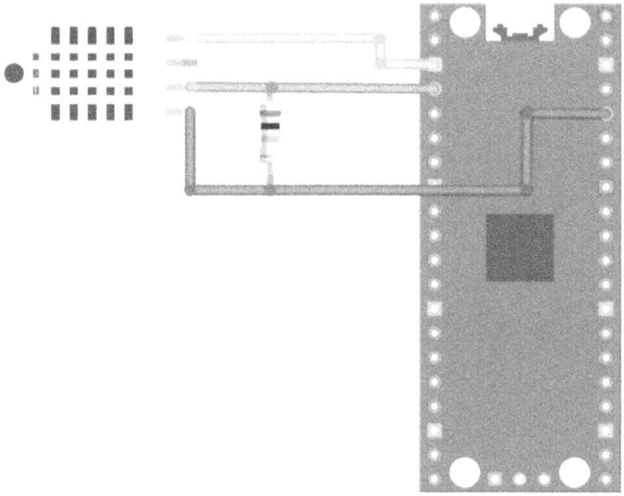

Pico	DHT22
3.5V OUT pin 36	VDD pin 1
GP2 pin 4	SDA serial data pin 2
GND pin 3	GND pin 4

It is very easy to create this circuit using a prototyping board and some jumper wires. You can also put the resistor close to the DHT22 to make a sensor package connected to the Pico using three cables.

The Software

With the hardware shown above connected to the Pico, the first thing that we need to do is establish that the system is working. The simplest way to do this is to pull the line down for 1ms and see if the device responds with a stream of pulses. These can be seen on a logic analyzer or an oscilloscope, both are indispensable tools. If you don't have access to either tool then you will just have to skip to the next stage and see if you can read in some data.

The simplest code that will do the job is:

```
from machine import Pin
from utime import sleep_ms

DHT=Pin(2,mode=Pin.OUT)
DHT.high
sleep_ms(1)
DHT.low()
sleep_ms(1)
DHT.init(mode=Pin.IN)
```

Setting the line initially high, to ensure that it is configured as an output, we then set it low, wait for around 1ms and then change its direction to input and so allow the line to be pulled high. If you run this as a "main" program then the change of direction takes about 1ms and so we don't need an additional sleep function call. If it is run as a method then it is much faster and you do need the call to sleep. It isn't easy to predict how fast MicroPython will run your code.

There is no need to set the line's pull-up mode because the Pico is the only device driving the line until it releases the line by changing direction to input. When a line is in input mode it is high impedance and this is why we need an external pull-up resistor in the circuit.

As long as the circuit has been correctly assembled and you have a working device, you should see something like:

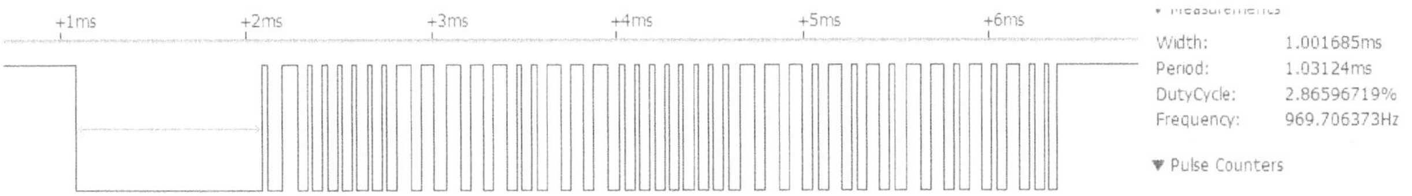

Reading the Data

With preliminary flight checks complete, it is time to read the 40-bit data stream. The first thing to do is wait for the low that the device sends before the start bit and then wait for the start bit:

```
for i in range(2):
    while DHT.value() == 1:
        pass
    while DHT.value() == 0:
        pass
```

The `for` loop waits for two falling edges.

Next we can start to read in the data. A total of 40 bits, i.e. 5 bytes, is difficult to work with. A good compromise is to read in the first 32 bits into a 32-bit integer and then read the final byte into a separate variable. The reason is that the fifth byte is a checksum, so we have separated out the data and the checksum, but there are many different ways to organize this task.
First we read the 32 data bits:

```
data = 0
t1 = ticks_us()
for i in range(32):
    while DHT.value() == 1:
        pass
    while DHT.value() == 0:
        pass
    t2 = ticks_us()
    data = data << 1
    data = data | ((t2 - t1) > 100)
    t1 = t2
```

You can see the general idea is to simply find the time between falling edges and then treat anything bigger then 100μs as a `1`. In practice, the measured times between falling edges is 80μs for a `0` and 120μs and `100` is a threshold halfway between the two. The bits are shifted into the variable data so that the first byte transmitted is the high-order byte.

Next we need to read the checksum byte:

```
checksum=0
for i in range(8):
    while DHT.value() == 1:
        pass
    while DHT.value() == 0:
        pass
    t2 = ticks_us()
    checksum = checksum << 1
    checksum = checksum |((t2 - t1) > 100)
    t1 = t2
```

This works in the same way. At the end of this we have 32 data bits in data and eight checksum bits in checksum and all we have to do is process the data to get the temperature and humidity.

Extracting the Data

You can process the data without unpacking it into individual bytes, but it is easier to see what is happening if we do:

```
byte1 = (data >> 24 & 0xFF)
byte2 = (data >> 16 & 0xFF)
byte3 = (data >> 8 & 0xFF)
byte4 = (data & 0xFF)
```

The first two bytes are the humidity measurement and the second two the temperature.

The checksum is just the sum of the first four bytes reduced to eight bits and we can test it using:

```
print("Checksum",checksum,(byte1+byte2+byte3+byte4)&0xFF)
```

If the two values are different, there has been a transmission error. The addition of the bytes is done as a full integer and then it is reduced back to a single byte by the AND operation. If there is a checksum error, the simplest thing to do is get another reading from the device. Notice, however, that you shouldn't read the device more than once every two seconds.

The humidity and temperature data are also easy to reconstruct as they are transmitted high byte first and 10 times the actual value.

Extracting the humidity data is easy:

```
humidity = ((byte1 <<8)| byte2) / 10.0
print("Humidity= ", humidity)
```

The temperature data is slightly more difficult in that the topmost bit is used to indicate a negative temperature. This means we have to test for the most significant bit and flip the sign of the temperature if it is set:

```
neg = byte3 & 0x80
byte3 = byte3 & 0x7F
temperature =(byte3 << 8 | byte4) / 10.0
if neg > 0:
    temperature = -temperature
print("Temperature=", temperature)
```

This completes the data processing.

The program as presented works, but it would benefit from refactoring into an object with suitable methods. A complete, refactored, listing including a main program can be seen below. This is just one way to break the program down into functions and exactly how best to do it depends on many factors.

```
from machine import Pin
from utime import sleep_ms, ticks_us


class DHT22():
    def __init__(self,gpio):
        self.pin=gpio
        self.pin = Pin(2, mode=Pin.OUT)
        self.pin.high()
        self.checksum=0
        self.temperature=0
        self.humidity=0
        sleep_ms(1)

    def getReading(self):
        DHT=self.pin
        DHT.low()
        sleep_ms(1)
        DHT.init(mode=Pin.IN)
        for i in range(2):
            while DHT.value() == 1:
                pass
            while DHT.value() == 0:
                pass

        data = 0
        t1 = ticks_us()
        for i in range(32):
            while DHT.value() == 1:
                pass
            while DHT.value() == 0:
                pass
            t2 = ticks_us()
            data = data << 1
            data = data | ((t2 - t1) > 100)
            t1 = t2
```

```
        checksum=0
        for i in range(8):
            while DHT.value() == 1:
                pass
            while DHT.value() == 0:
                pass
            t2 = ticks_us()
            checksum = checksum << 1
            checksum = checksum |((t2 - t1) > 100)
            t1 = t2
        byte1 = (data >> 24 & 0xFF)
        byte2 = (data >> 16 & 0xFF)
        byte3 = (data >> 8 & 0xFF)
        byte4 = (data & 0xFF)
        self.checksum=(checksum ==(byte1+byte2+byte3+byte4)&0xFF)
        self.humidity = ((byte1 <<8)| byte2) / 10.0
        neg = byte3 & 0x80
        byte3 = byte3 & 0x7F
        self.temperature =(byte3 << 8 | byte4) / 10.0
        if neg > 0:
            self.temperature = -self.temperature
```

Once you have the class you can write simple programs like:

```
dht=DHT22(2)
dht.getReading()
print("Checksum",dht.checksum)
print("Humidity= ", dht.humidity)
print("Temperature=", dht.temperature)
```

For real use, the program has the drawback that if the DHT22 fails to respond the `getReading` method will hang. The solution is to add a counter to the `while` loops to provide a timeout after so many attempts.

DHT22 Using the PIO – Counting

The DHT22 is not well suited to working with the PIO. The reason is that its protocol doesn't use fixed time slots for each bit or synchronization via a separate clock, two situations where it is usually possible to write compact PIO programs to send and receive data. The DHT22 relies on the length of the slot used to transmit a single bit to code zero or one and this means that decoding is a matter of measuring the time between edges. The PIO doesn't have access to a timer and implementing an instruction clock counter isn't straightforward – but it can be done. The following PIO program does exactly this, but it is a long program and it only just satisfies the constraints placed on it to work. After presenting the obvious algorithm with a slightly difficult implementation, a simpler and more direct but less obvious method is implemented.

Before going into details, it is worth explaining what the constraints are and what makes PIO access worth the effort. The bit-banging solution given earlier works and is perfectly good. However, it demands the full attention of the processor while the data is being read in. If an interrupt occurs during reading, the data will be corrupt. On the other hand, if you can implement a PIO program that reads in the data the processor is free to do something else until it wants to use the data and interrupts have no effect. This all depends on the PIO being able to read the data from the device and keep it stored ready for when the processor wants to read it. If you consider this for a few moments, it means that you can hold data in the 4-word FIFO buffer and an additional word in the ISR, making a total of 160 bits of temporary storage, which seems to be more than enough for the 40 bits that the DHT22 produces. However, for a timing-based protocol, things are more complicated.

We can write a loop that counts instruction cycles when the input line is high and this count can be up to 32 bits in resolution. However, there is no easy way to convert the 32-bit count into single bits by applying a threshold as in the bit-banging example. What this means is that we are forced to send the bit counts, rather than the decoded bits back to the processor. A little arithmetic reveals that we can store the counts for all 40 bits, as long as the count is represented as a 4-bit number. We can store the first 32-bits, the data in the 4-word FIFO buffer with the counts for each byte, in a single element. The final byte can be stored in the ISR until the processor reads the FIFO register and frees up space. A count of 4 bits is enough to tell the difference between the times for a zero and a one and this makes the whole scheme possible, as long as we adjust the clock frequency correctly.

What all this means is that we can write a PIO program that can be set to read the data from the DHT22 and store the result in the FIFO and the ISR until the processor is ready to read it. The downside is that the processor has to convert the timing counts to bits and pack the results into bytes before processing the humidity, temperature and checksum – a small price to pay.

With all this worked out we can write the PIO program. The first problem is that we have to send a 1ms initialization pulse:

```
.program dht
       set pins, 1
again:
       pull block
       set pins, 0
       mov x, osr
loop1:
       jmp x--,loop1
       set pindirs, 0
```

This uses a data value written to the state machine by a MicroPython program to execute a delay loop that takes 1ms. It also serves to start the conversion. Also notice that the low pulse is terminated by setting the pin to input using `set pindirs`.

Next we have to wait for two rising edges to pass before we get to the data pulses:

```
        wait 1 pin 0
        wait 0 pin 0
        wait 1 pin 0
        wait 0 pin 0
```

Finally we can start reading in the four bytes of data:

```
        set y,31
bits:
        wait 1 pin 0
        set x, 0
loop2:
        jmp x--,continue
continue: jmp pin,loop2
        in x,4
        jmp y--,bits
```

The `bits` loop processes all 32 bits. The inner `loop2` processes each bit in turn. The `x` register is set to zero and then decremented using a forward jump to the end of the loop. Notice that it doesn't matter if the jump is taken, the next instruction is always `continue`. Decrementing the `x` register when it is already at zero seems to increment it, but as this is not part of the PIO specification there is a small danger that this behavior might change. Once the loop has completed, the 4-bit count is transferred to the ISR and the next bit starts. Notice that it is assumed that auto-push is set so that the FIFO is automatically added to when the ISR is full.

We have to repeat the process for the checksum:

```
        set y,7
check:
        wait 1 pin 0
        set x, 0
loop3:
        jmp x--,continue2
continue2: jmp pin,loop3
        in x,4
        jmp y--,check
jmp again
```

Notice that if the MicroPython program hasn't read the FIFO then this blocks and waits for free space. Finally, we jump back to the start of the PIO program to wait for the MicroPython program to write to the FIFO.

The MicroPython program to make all this work is fairly straightforward, but it involves a lot of bit manipulation to convert the counts into bits:

```
    def getByte(self):
        count = self.sm.get()
        byte = 0
        for i in range(8):
            byte = byte << 1
            if ((count >> i * 4) & 0x0F) > 8:
                byte = byte | 1
        return byte
```

This reads a word from the PIO and converts it into a byte by comparing each 4-bit count with a threshold, i.e. 8. We also need to set the shift direction to a right shift to keep the bits the correct order.

Using this function we can obtain the five bytes needed to compute the humidity, temperature and checksum as per the previous program. This makes it very easy to convert the DHT22 class to use the PIO program.

The complete program to read a DHT22 is:

```
import rp2
from machine import Pin

@rp2.asm_pio(set_init=rp2.PIO.OUT_LOW, autopush=True,
in_shiftdir=rp2.PIO.SHIFT_RIGHT)
def dht22():
    wrap_target()
    label("again")
    pull(block)
    set(pins, 0)
    mov(x, osr)
    label("loop1")
    jmp(x_dec, "loop1")
    set(pindirs, 0)

    wait(1, pin, 0)
    wait(0, pin, 0)
    wait(1, pin, 0)
    wait(0, pin, 0)
```

```
    set(y, 31)
    label("bits")
    wait(1, pin, 0)
    set(x, 0)
    label("loop2")
    jmp(x_dec, "continue")
    label("continue")
    jmp(pin, "loop2")
    in_(x, 4)
    jmp(y_dec, "bits")

    set(y, 7)
    label("check")
    wait(1, pin, 0)
    set(x, 0)
    label("loop3")
    jmp(x_dec, "continue2")
    label("continue2")
    jmp(pin, "loop3")
    in_(x, 4)
    jmp(y_dec, "check")
    wrap()


class DHT22():
    def __init__(self, gpio):
        self.sm = rp2.StateMachine(0, dht22,
                        freq=976562, in_base=Pin(gpio),
                         set_base=Pin(gpio), jmp_pin=Pin(gpio))
        self.sm.active(1)

    def getByte(self):
        count = self.sm.get()
        byte = 0
        for i in range(8):
            byte = byte << 1
            if ((count >> i * 4) & 0x0F) > 8:
                byte = byte | 1
        return byte
```

```
    def getReading(self):
        self.sm.put(1000)
        byte1 = self.getByte()
        print(byte1)
        byte2 = self.getByte()
        byte3 = self.getByte()
        print(hex(byte2))
        byte4 = self.getByte()
        checksum = self.getByte()
        self.checksum =
                 (checksum == (byte1+byte2+byte3+byte4) & 0xFF)
        self.humidity = ((byte1 << 8) | byte2) / 10.0
        neg = byte3 & 0x80
        byte3 = byte3 & 0x7F
        self.temperature = (byte3 << 8 | byte4) / 10.0
        if neg > 0:
            self.temperature = -self.temperature


dht = DHT22(2)
dht.getReading()
print("Checksum", dht.checksum)
print("Humidity= ", dht.humidity)
print("Temperature=", dht.temperature)
```

The initialization is the only function of additional interest as most of the rest of the program follows the previous example. It sets the clock frequency so that instructions take $1\mu s$. This gives a count in the region of 4 to 13 and the threshold is set accordingly. At this clock speed, you also need to set the `x` register to `1000` to get the 1ms initialization pulse. It also sets the auto-push to 32 bits and a right shift. Notice that, as well as the SET and IN groups, you have to set the jump pin to the GPIO line being used or the counting loop doesn't work.

DHT22 Using the PIO – Sampling

A simpler alternative way for decoding the data is to ignore the fact that it is the width of each bit's frame that defines a zero, a short frame, or a one a long frame, and notice that if you sample at a suitable fixed time from the rising edge of a pulse then you will get a zero in a zero frame and a one in a one frame:

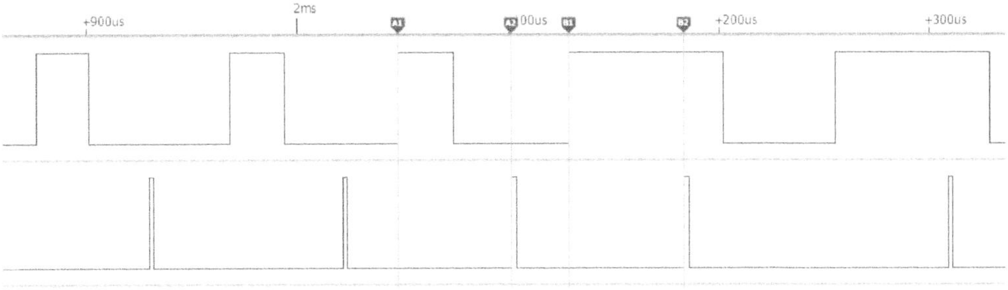

You can see the sampling times from the pulses on the lower trace of the logic analyzer and the fact that you do indeed get a `0` in a zero frame and a `1` in a one frame. You can also see that the time to sample from the rising edge is constant, even if the sampling period varies.

Surprisingly, the PIO program to implement this sampling isn't significantly simpler than the previous program, but the entire four data bytes can now be packed into a single FIFO entry and the checksum byte can occupy a second entry.

The PIO program starts off in the same way:

```
.program dht
    set pins, 1
    set pindirs, 1
again:
  pull block
  set pins, 0
mov x, osr
loop1:
    jmp x--,loop1
set pindirs, 0
wait 1 pin 0
wait 0 pin 0
wait 1 pin 0
wait 0 pin 0
```

The waits' position is at the state of the data and now we can wait for a rising edge and then delay until the sample time:

```
set y,31
bits:
   wait 1 pin 0 [25]
    in pins,1
    wait 0 pin 0
 jmp y--,bits
```

This too is very similar, only now we don't keep a count but simply add the single bit sampled from the input line to the ISR. When all 32 bits have been added, the `autopush` moves the data to the FIFO.

The eight checksum bits are just as easy to process and the complete PIO program is:

```
.program dht
 set pins, 1
 set pindirs, 1
again:
 pull block
 set pins, 0
mov x, osr
loop1:
 jmp x--,loop1
set pindirs, 0
wait 1 pin 0
wait 0 pin 0
wait 1 pin 0
wait 0 pin 0

set y,31
bits:
 wait 1 pin 0 [25]
 in pins,1
 wait 0 pin 0
 jmp y--,bits

 set y,7
check:
wait 1 pin 0 [25]
 in pins,1
 wait 0 pin 0
 jmp y--,check
push block
jmp again
```

Notice that we have to manually push the ISR because it doesn't fill up just the 8 bits of the checksum.

The MicroPython program is also very like the previous program, but we have to change the clock frequency and the shift direction to make decoding the bytes from the 32-bit word easier.

We need a slower clock speed to make the delay position the sample point about 50μs from the rising edge – recall that the delay is limited to 32 clock cycles.

The `getReading` method now might as well read the FIFO and decode the data directly as it is a much simpler task:

```
    def getReading(self):
        self.sm.put(500)
        data=0
        data = self.sm.get()
        byte1 = (data >> 24 & 0xFF)
        byte2 = (data >> 16 & 0xFF)
        byte3 = (data >> 8 & 0xFF)
        byte4 = (data & 0xFF)
        checksum = self.sm.get() & 0xFF
```

Notice the need to change the value sent to the input to account for the slower clock rate – `500` now produces a 1ms initialization pulse. The rest of the program is unchanged as we now have the four data bytes and the checksum as before. This program is simpler to implement and can be extended to a protocol that needs to read in more than 40 bits at a time.

Complete Listing

```
import rp2
from machine import Pin

@rp2.asm_pio(set_init=(rp2.PIO.OUT_LOW,rp2.PIO.OUT_LOW),
             autopush=True, in_shiftdir=rp2.PIO.SHIFT_LEFT)
def dht22():
    wrap_target()
    label("again")
    pull(block)
    set(pins, 0)
    mov(x, osr)
    label("loop1")
    jmp(x_dec, "loop1")
    set(pindirs, 0)

    wait(1, pin, 0)
    wait(0, pin, 0)
    wait(1, pin, 0)
    wait(0, pin, 0)
```

```
    set(y, 31)
    label("bits")
    wait(1, pin, 0) [25]
    in_(pins, 1)
    wait(0, pin, 0)
    jmp(y_dec, "bits")

    set(y, 7)
    label("check")
    wait(1, pin, 0)[25]
    set(pins,2)
    set(pins,0)
    in_(pins, 1)
    wait(0, pin, 0)
    jmp(y_dec, "check")
    push(block)
    wrap()


class DHT22():
    def __init__(self, gpio):
        self.sm = rp2.StateMachine(0, dht22, freq=490196,
           in_base=Pin(gpio), set_base=Pin(gpio), jmp_pin=Pin(gpio))
        self.sm.active(1)

    def getReading(self):
        self.sm.put(500)
        data=0
        data = self.sm.get()
        byte1 = (data >> 24 & 0xFF)
        byte2 = (data >> 16 & 0xFF)
        byte3 = (data >> 8 & 0xFF)
        byte4 = (data & 0xFF)
        checksum = self.sm.get() & 0xFF
        self.checksum =
            (checksum == (byte1+byte2+byte3+byte4) & 0xFF)
        self.humidity = ((byte1 << 8) | byte2) / 10.0
        neg = byte3 & 0x80
        byte3 = byte3 & 0x7F
        self.temperature = (byte3 << 8 | byte4) / 10.0
        if neg > 0:
            self.temperature = -self.temperature


dht = DHT22(2)
dht.getReading()
print("Checksum", dht.checksum)
print("Humidity= ", dht.humidity)
print("Temperature=", dht.temperature)
```

Summary

- The DHT22 is a low-cost temperature and humidity sensor.
- It uses a custom single wire bus which is not compatible with the 1-Wire bus.
- Its asynchronous protocol is easy to implement directly in user space.
- A very simple checksum is used to detect errors.
- It is possible to implement the protocol as defined in the datasheet using a PIO by using counting loops to time each pulse.
- A better use of the PIO is to notice that the protocol can be decoded by testing the state of the line a fixed time after the rising edge.

Chapter 15
The 1-Wire Bus And The DS1820

The 1-Wire bus is a proprietary bus that is very easy to use and has a lot of useful devices you can connect to it, including the iButton security devices, memory, data loggers, fuel gauges and more. However, probably the most popular of all 1-Wire devices is the DS18B20 temperature sensor - it is small, very cheap and very easy to use. This chapter show you how to work with it, but first let's deal with the general techniques needed to work with the 1-Wire bus.

Programming in MicroPython on the Pico only just has enough speed to implement a bit-banging driver, which, even so, is enough for many purposes. However, the 1-Wire protocol is easy enough to implement using the PIO. In this chapter we look at the basic ideas of the 1-Wire bus, implement a bit banging program for its most popular device and extend this to a PIO program.

The Hardware

One-wire devices are very simple and only use a single wire to transmit data:

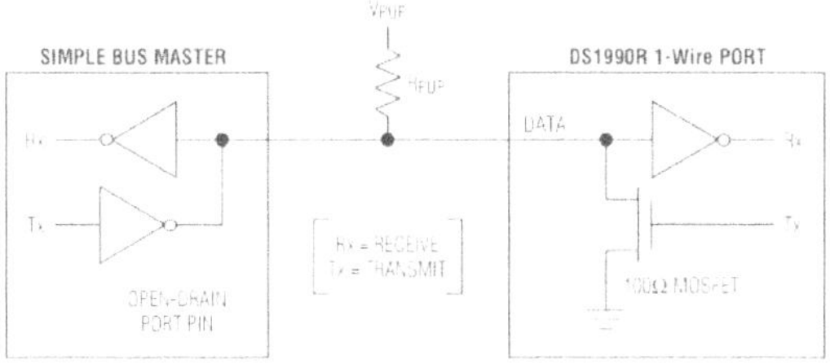

The 1-Wire device can pull the bus low using its Tx line and can read the line using its Rx line. The reason for the pull-up resistor is that both the bus master and the slave can pull the bus low and it will stay low until they both release the bus.

The device can even be powered from the bus line by drawing sufficient current through the pull-up resistor - so called parasitic mode. Low power devices work well in parasitic mode, but some devices have such a heavy current draw that the master has to provide a way to connect them to the power line - so called strong pull-up. In practice, parasitic mode can be difficult to make work reliably for high-power devices.

In normal-powered mode there are just three connections - V power, usually 3.3V for the Pico, Ground and Data:

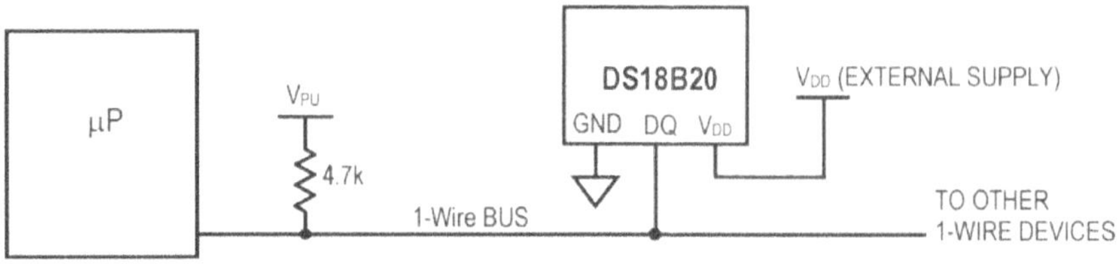

The pull-up resistor varies according to the device, but anything from 2.2K to 4.7kΩ works. The longer the bus, the lower the pull-up resistor has to be to reduce "ringing". There can be multiple devices on the bus and each one has a unique 64-bit lasered ROM code, which can be used as an address to select the active devices.

For simplicity, it is better to start off with a single device and avoid the problem of enumerating the devices on the bus, although once you know how everything works this isn't difficult to implement. To get started, select a 1-wire device that you want to work with and set it up ready to talk to the Pico, in this case the DS18B20 is explained. The functions described in this chapter should work with any 1-wire device.

Initialization

Every transaction with a 1-wire device starts with an initialization handshake. First we have to work out how to configure the GPIO line. This example assumes that the 1-wire device is connected to GP2.

You might think that we have to initialize the GPIO line so that it works in pull-up mode. This isn't necessary and the default push-pull mode will do. The reason is that, in the case of the 1-wire bus, the master controls when other devices send their data. Typically the master sends a pulse and then the slaves respond by pulling the line low. As long as the master doesn't drive the line during the period when the slaves are responding, everything is fine.

What we do in practice is to configure the GPIO line for output only when the master needs to drive the line. Once the master is finished the GPIO line is set back to input and the pull-up resistor is allowed to pull the line back up. After this, any slave wanting to send data is free to pull the line low.

The first transaction we need is the initialization pulse. This is simply a low pulse that lasts at least 480μs, a pause of 15μs to 60μs follows and then any and all of the devices on the bus pull the line low for 60μs to 240μs.

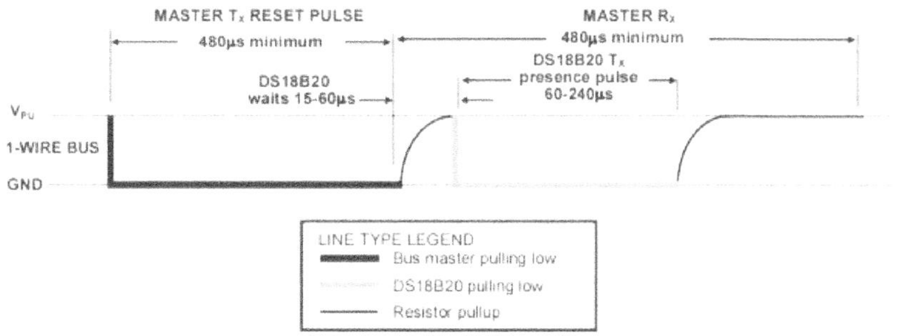

The suggested timings are set the line low for 480μs and read the line after 70μs followed by a pause of 410μs.

This is fairly easy to implement as a function:

```
def presence(pin):
    pin.init(mode=Pin.OUT)
    pin.high()
    sleep_ms(1)
    pin.low()
    sleep_us(480)
    pin.init(mode=Pin.IN)
    sleep_us(70)
    b = pin.value()
    sleep_us(410)
    return b
```

We pull the line low for 480μs and then let it be pulled back up by changing the line to input, i.e. high impedance. After a 70μs wait, which is right at the start of the guaranteed period when the line should be low if there is an active device on the bus, we read the input line and then wait another 410μs to complete the data slot.

The timings in this case are not critical as long as the line is read while it is held low by the slaves, which is never less than 60μs and is typically as much as 100μs. If there is a device, the function should return a `0` and if there are no devices it should return a `1`.

```
pin=Pin(2,Pin.OUT)
pin.high()
if presence(pin) == 1:
    print("No device")
else:
    print("Device present")
```

If you try this partial program and have a logic analyzer with a 1-wire protocol analyzer you will see something like:

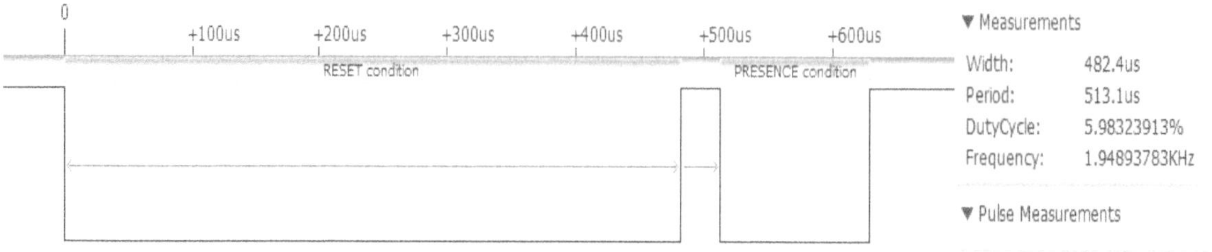

Seeing a presence pulse is the simplest and quickest way to be sure that your hardware is working.

Writing Bits

Our next task is to implement the sending of some data bits to the device. The 1-Wire bus has a very simple data protocol. All bits are sent using a minimum of 60μs for a read/write slot. Each slot must be separated from the next by a minimum of 1μs.

The good news is that timing is only critical within each slot. You can send the first bit in a slot and then take your time before you send the next bit as the device will wait for you. This means you only have to worry about timing within the functions that read and write individual bits.

To send a 0 you have to hold the line low for most of the slot. To send a 1 you have to hold the line low for just between 1μs and 15μs and leave the line high for the rest of the slot. The exact timings can be seen below:

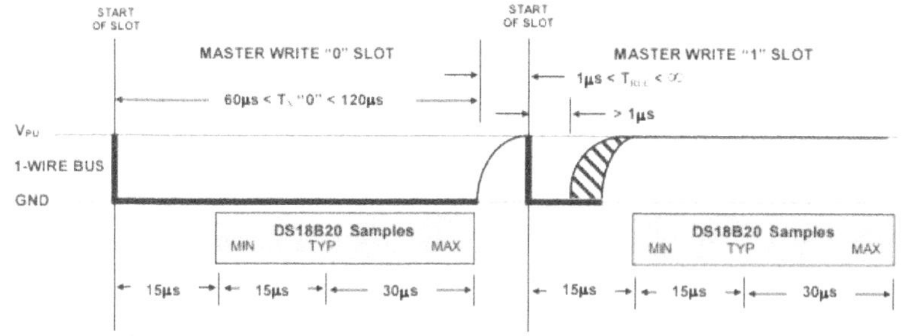

It seems reasonable to use the typical timings given in the datasheets. So for a `0` we hold the line low for 60μs then let it go high for the remainder of the slot, 10μs. To send a `1` we hold the line for 6μs and then let it go high for the remainder of the slot, 64μs. As the only time critical operations are the actual setting of the line low and then back to high, there is no need to worry too much about the speed of operation of the entire function so we might as well combine writing `0` and `1` into a single `writeBit` function. The problem is that these timings are very close to what is possible using MicroPython. To get the necessary speed and accuracy we need to use busy wait `for` loops for the timing and, while the function does just work without, it benefits from using native compilation:

```
@micropython.native
def writeBit(pin,b):
    if b == 1:
        delay1 = 1
        delay2 = 30
    else:
        delay1 = 30
        delay2 = 0
    pin.low()
    for i in range(delay1):
        pass
    pin.high()
    for i in range(delay2):
        pass
```

The code at the start of the function simply increases the time between slots slightly. To avoid having to set the line's output mode between each bit, it is assumed that any function that uses `writeBit` sets the line to OUT before calling it and to IN after calling it. This allows the line to be pulled high, ready for any response from the slave after a set of bits has been sent.

You can see two ones followed by two zeros in the following logic analyzer trace:

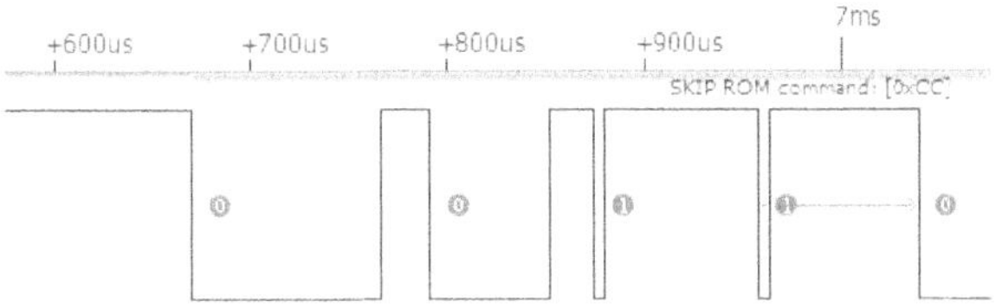

The initial pulse for a `1` at $6\mu s$ is just short enough to work. If the function is not compiled to native code then it is closer to $12\mu s$ and, while it tends to work, it is more prone to error.

A First Command - Writing Bytes

After discovering that there is at least one device connected to the bus, the master has to issue a ROM command. In many cases the ROM command used first will be the Search ROM command, which enumerates the 64-bit codes of all of the devices on the bus. After collecting all of these codes, the master can use Match ROM commands with a specific 64-bit code to select the device the master wants to talk to.

While it is perfectly possible to implement the Search ROM procedure, it is simpler to work with the single device by using commands which ignore the 64-bit code and address all of the devices on the bus at the same time. Of course, this only works as long as there is only one device on the bus. If there is only one device then we can use the Skip ROM command `0xCC` to tell all the devices on the bus to be active.

We now need a function that can send a byte. As we have a `writeBit` function this is easy:

```
def sendskip(pin):
        writeBit(pin, 0)
        writeBit(pin, 0)
        writeBit(pin, 1)
        writeBit(pin, 1)
        writeBit(pin, 0)
        writeBit(pin, 0)
        writeBit(pin, 1)
        writeBit(pin, 1)
```

Notice that `0xCC` is `1100 1100` in binary and the 1-Wire bus sends the least significant bit first. If you try this out you should find it works, but the device doesn't respond because it is waiting for another command. Again, as the time between writing bits isn't critical, we can take this first implementation of the function and write something more general, if slightly slower.

The `writeByte` function will write the low eight bits of an `int` to the device:

```
def writeByte(pin, byte):
    pin.init(mode=Pin.OUT)
    for i in range(8):
        writeBit(pin, byte & 1)
        byte = byte >> 1
    pin.init(mode=Pin.IN)
```

Using this we can send a Skip ROM command using:

```
writeByte(2, 0xCC)
```

You can see the pattern of bits sent on a logic analyzer:

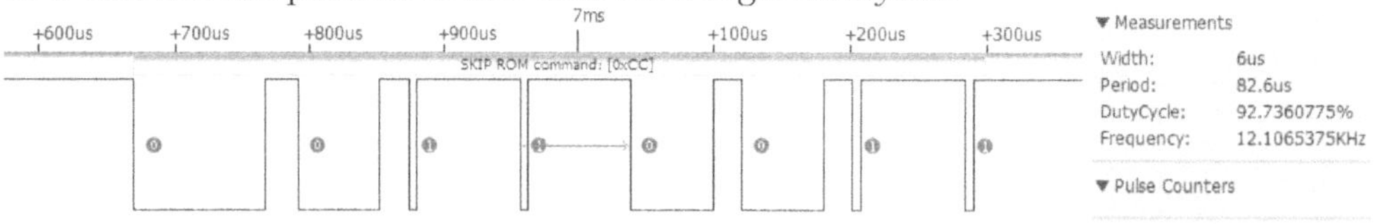

Reading Bits

We already know how the master sends a `1` and a `0`. The protocol for the slave device is exactly the same except that the master still provides the slot's starting pulse. That is, the master starts a 60μs slot by pulling the bus down for at least 1μs. Then the slave device either holds the line down for a further 15μs minimum or it simply allows the line to float high. See below for the exact timings:

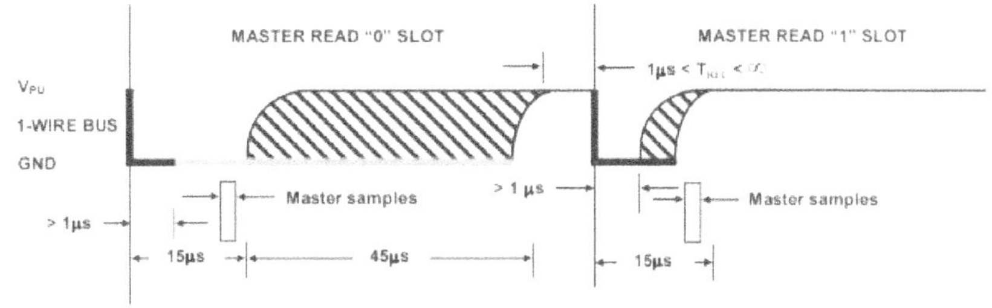

So all we have to do to read bits is to pull the line down for more than 1μs and then sample the bus after pausing long enough for the line to be pulled up or held low. The datasheet gives 6μs for the master's pulse and a 9μs pause. In practice, even native code is only just fast enough as the time to change from output to input is quite high and we need to read the input line at once to avoid missing the data:

```
@micropython.native
def readBit(pin):
    pin.init(mode=Pin.OUT)
    pin.low()
    pin.high()
    pin.init(mode=Pin.IN)
    b = pin.value()
    sleep_us(60)
    return b
```

A logic analyzer shows the typical pattern of bits from the device:

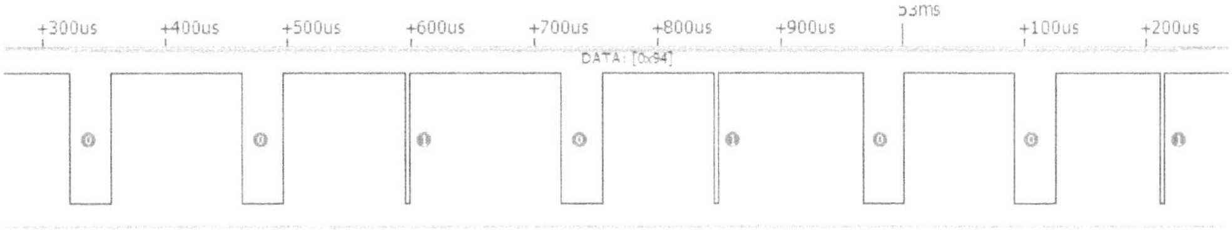

By adding some commands to toggle a line after the sample is taken, we can see how the timing works:

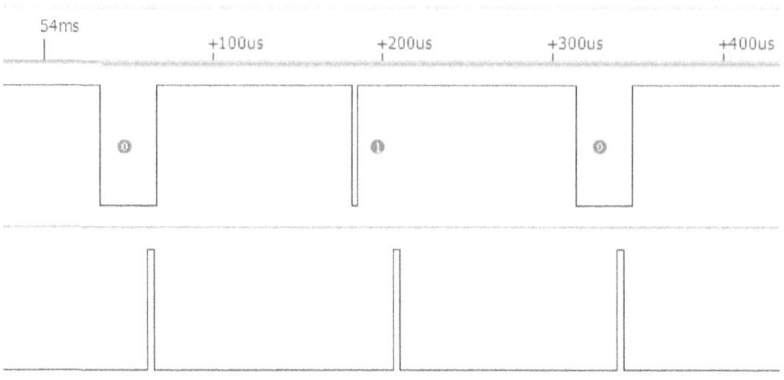

The reading is taken just before the indicator pulse on the lower trace. You can see that it is within the timing window to read a zero or a one correctly, but it would benefit from being closer to the down going edge.

Finally, we need a function that will read a byte. As in the case of writing a byte, there is no time criticality in the time between reading bits, so we don't need to take extra special care in constructing the function:

```
def readByte(pin):
    byte = 0
    for i in range(8):
        byte = byte | readBit(pin) << i
    return byte
```

The only difficult part is to remember that the 1-Wire bus sends the least significant bit first and so this has to be shifted into the result from the right.

The sequence is:

- Test to see if a device is present:
  `presence(uint8_t pin)`
- Write a byte:
  `writeByte(pin, byte)`
- Read a byte:
  `readByte(pin)`

These functions will be used in the rest of the chapter to work with real 1-wire devices.

These are not the only functions we need to work with the 1-Wire bus. We need to be able to compute the CRC error checks that are commonly used to confirm that data has been transmitted correctly and we need to perform a ROM search to discover what devices are connected to the bus, but this is beyond the scope of this introduction. For details of a general C implementation, see *Raspberry Pi IOT in C, Second Edition*, ISBN:9781871962635.

Computing The CRC

We have already encountered the idea and implementation of a CRC (Cyclic Redundancy Checksum) in Chapter 12. The 1-Wire bus uses the same CRC for all its devices and therefore we need to implement it just once. This is perhaps not in the most efficient way, but it will work. For low data rate applications high efficiency isn't needed and you can make use of a direct implementation. The 1-Wire datasheet specifies the CRC used in 1-wire devices as a shift register rather than as a polynomial equation:

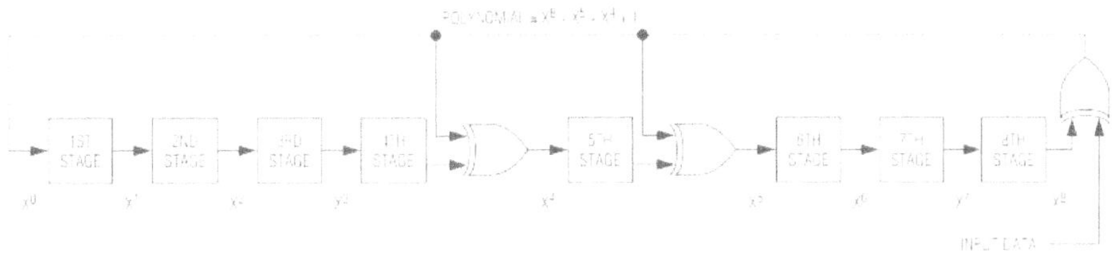

However, this is equivalent to a generator polynomial that defines the CRC as it is simply the hardware implementation of the calculation.

In this case it is:

$X^8 + X^5 + X^4 + 1$

The first question to answer is, what is the connection between binary values, polynomials and shift-registers? The answer is that you can treat a binary number as the coefficients of a polynomial, for example `101` is $1*X^2+0*X+1$. Each bit position corresponds to a power of X. Using this notation creates a very simple relationship between multiplying by X and a left-shift.

For example:

$(1*X^2 + 0*X+ 1)*X = 1*X^3 + 0*X^2 + 1X + 0$

corresponds to:

```
101 <<1 == 1010
```

You can see that this extends to multiplying one polynomial by another and even polynomial division, all accomplished by shifting and XOR (eXclusive OR).

The CRC is the remainder when you divide the polynomial that represents the data by the generator polynomial. The computation of the remainder is what the shift register specified on the datasheet does. The fact that the division can be implemented so simply in hardware is what makes this sort of CRC computation so common. All the hardware has to do is zero the shift register and feed the data into it. When all the data has been shifted in, what is left in the shift register is the CRC, i.e. the remainder.

To check the data you have received, all you have to do is run it through the shift register again and compare the computed CRC with the one received. A better trick is also to run the received CRC through the shift register. If there have been no errors, this will result in `0`.

You can look into the theory of CRCs, bit sequences and polynomials further, it is interesting and practically useful, but we now know everything we need to if we want to implement the CRC used by1-Wire devices. All we have to do is implement the shift register in software.

From the diagram, what we have to do is take each bit of the input data and XOR it with the least significant bit of the current shift register. If the input bit is `0`, the XORs in the shift register don't have any effect and the CRC just has to be moved one bit to the right. If the input bit is `1`, we have to XOR the bits at positions 3 and 4 with `1` and put a `1` in at position 7 to simulate shifting a `1` into the register, i.e. XOR the shift register with `10001100`.
So the algorithm for a single byte is:

```
for j in range(8):
    temp = (crc ^ databyte) & 0x01
    crc >>= 1
    if temp:
       crc ^= 0x8C
    databyte >>= 1
return crc
```

First we XOR the data with the current CRC and extract the low-order bit into `temp`. Then we right-shift the CRC by one place. If the low-order result stored in temp was a `1`, we have to XOR the CRC with `0x8C` to simulate the XORs in the shift register and shift in a `1` at the most significant bit. Then shift the data one place right and repeat for the next data bit.

With this worked out, we can now write a `crc8` function that computes the CRC for the entire eight bytes of data:

```
def crc8(data,len):
    crc = 0
    for i in range(len):
        databyte = data[i]
        for j in range(8):
            temp = (crc ^ databyte) & 0x01
            crc >>= 1
            if temp:
                crc ^= 0x8C
            databyte >>= 1
    return crc
```

With this in place we can now check the CRC of any data a 1-Wire bus device sends us.

The DS18B20 Hardware

Now let's put all of this to work with the most popular 1-Wire device, the DS18B20. It is available in a number of formats, but the most common makes it look just like a standard BJT (Bipolar Junction Transistor) which can sometimes be a problem when you are trying to find one. You can also get them made up into waterproof sensors complete with cable.

No matter how packaged, they will work at 3.3V or 5V.

The basic specification of the DS18B20 is:

- Measures temperatures from -55°C to +125°C (-67°F to +257°F)
- ±0.5°C accuracy from -10°C to +85°C
- Thermometer resolution is user-selectable from 9 to 12 bits
- Converts temperature to 12-bit digital word in 750ms (max)

It can also be powered from the data line, allowing the bus to operate with only two wires - data and ground. However, this "parasitic power" mode is difficult to make work reliably and best avoided in an initial design.

To supply it with enough power during a conversion, the host has to connect it directly to the data line by providing a "strong pull-up" - essentially a transistor. In normal-powered mode there are just three connections:

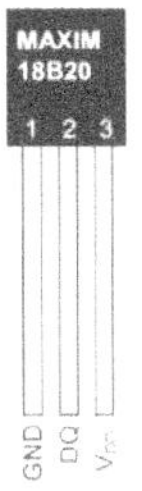

Ground needs to be connected to the system ground, VDD to 3.3V and DQ to the pull-up resistor of an open collector bus.

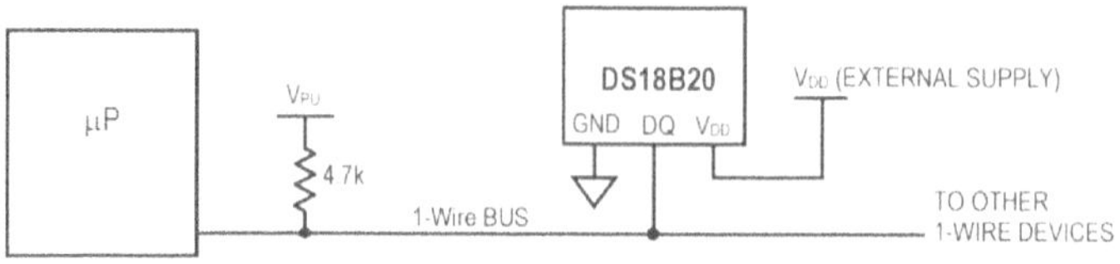

While you can have multiple devices on the same bus, for simplicity it is better to start off with a single device until you know that everything is working.

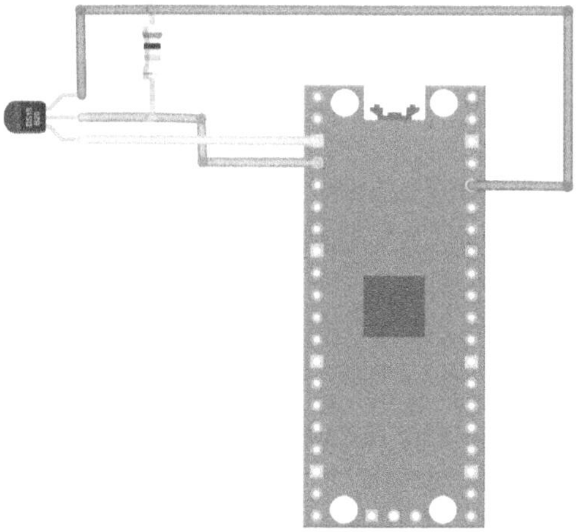

You can build the circuit in a variety of ways. You can solder the resistor to the temperature sensor and then use some longer wires with clips to connect to the Pico. You could also solder directly to the Pico or use a prototyping board.

Initialization

Every transaction with the a 1-wire device starts with an initialization handshake. The master holds the bus low for at least 480μs, a pause of 15μs to 60μs follows, and then any and all of the devices on the bus pull the line low for 60μs to 240μs.

For this we need the presence function:

```
    if presence(2) == 1:
        print("No device \n")
    read temperature
        ...
```

If you try this partial program and have a logic analyzer with a 1-wire protocol analyzer, you will see something like:

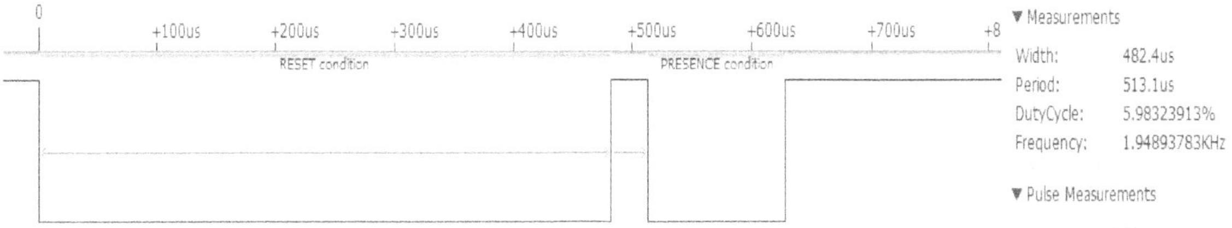

From this point it is just a matter of using the functions developed previously to work with the commands defined in the datasheet.

Initiating Temperature Conversion

As there is only one device on the 1-Wire bus, we can use the Skip ROM command (0xCC) to signal it to be active:

```
writeByte(pin, 0xCC)
```

You can see the pattern of bits sent on a logic analyzer:

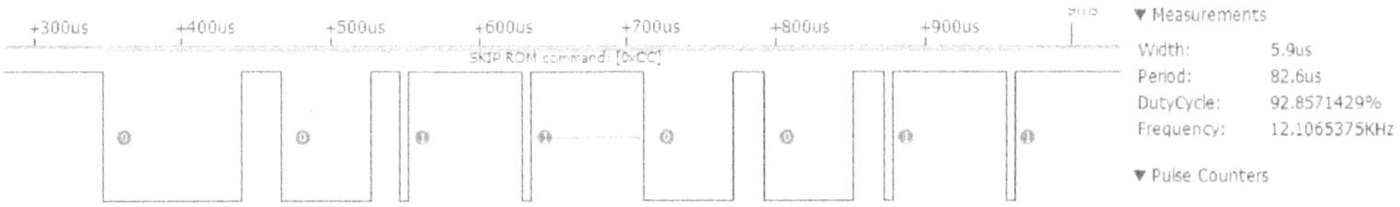

Our next task is to send a Convert command, 0x44, to start the DS18B20 making a temperature measurement. Depending on the resolution selected, this can take as long as 750ms. How the device tells the master that the measurement has completed depends on the mode in which it is operating, but using an external power line, i.e. not using parasitic mode, the device sends a 0 bit in response to a bit read until it is completed when it sends a 1.

This is how 1-Wire devices that need time to get data ready slow down the master until they are ready. The master can read a single bit as often as it likes and the slave will respond with a 0 until it is ready with the data.

As we already have a readBit function this is easy. The software polls for the completion by reading the bus until it gets a 1 bit:

```
def convert(pin):
    writeByte(pin, 0x44)
    for i in range(500):
        sleep_ms(10)
        if readBit(pin) == 1:
            j=i
            break
```

You can of course test the return value to check that the result has been obtained. If `convert` returns `500` then the loop times out. When the function returns, the new temperature measurement is stored in the device's scratchpad memory and now all we have to do is read this.

Reading the Scratchpad

The scratchpad memory has nine bytes of storage in total and does things like control the accuracy of conversion and provide status information. In our simple example the only two bytes of any great interest are the first two, which hold the result of a temperature conversion. However, as we are going to check the CRC for error detection, we need to read all nine bytes.

All we have to do is issue a Read Scratchpad, `0xBE`, command and then read the nine bytes that the device returns. To send the new command we have to issue a new initialization pulse and a Skip ROM, `0xCC`, command followed by a Read Scratchpad command, `0xBE`:

```
presence(pin)
writeByte(pin, 0xCC)
writeByte(pin, 0xBE)
```

Now the data is ready to read. We can read all nine bytes of it or just the first two that we are interested in. The device will keep track of which bytes have been read. If you come back later and read more bytes you will continue the read from where you left off. If you issue another initialization pulse then the device aborts the data transmission.

As we do want to check the CRC for errors, we will read all nine bytes:

```
data=[]
for i in range(9):
    data.append(readByte(pin))
```

Now we have all of the data stored in the scratchpad and the CRC byte, we can check for errors:

```
print(crc8(data,9))
```

As before, the value returned will be `0` if there are no transmission errors. The `crc8` function was given earlier.

Getting The Temperature

To obtain the temperature measurement we need to work with the first two bytes, which are the least and most significant bytes of the 12-bit temperature reading:

```
t1 = data[0]
t2 = data[1]
```

`t1` holds the low-order bits and `t2` the high-order bits.

All we now have to do is to put the two bytes together as a 16-bit two's complement integer. We can do this very easily:

```
temp1 = (t2 << 8 | t1)
```

As this is a 32-bit integer, you will have to propagate the sign bit manually into the high order bits to get a negative number:

```
if t2 & 0x80:
    temp1=temp1 | 0xFFFF0000
```

Finally, we have to convert the temperature to a scaled value. As the returned data gives the temperature in centigrade with the low-order four bits giving the fractional part, it has to be scaled by a factor of 1/16:

```
temp=temp1/16
print(temp)
```

A Temperature Class

Obviously once we have some working functions it is time to package everything into a suitable class:

```
class DS18B20:
    def __init__(self,pin):
        self.pin=Pin(pin,mode=Pin.IN)
        self.pin.high()
```

The other functions are simply converted into methods of the class. The only new method is:

```
    def getTemp(self):
        if self.presence()==1:
            return -1000
        self.writeByte(0xCC)
        if self.convert()==500:
            return -3000
        self.presence()
        self.writeByte( 0xCC)
        self.writeByte( 0xBE)
        data=[]
        for i in range(9):
            data.append(self.readByte())
        if self.crc8(data,9)!=0:
            return -2000
        t1 = data[0]
        t2 = data[1]
        temp1 = (t2 << 8 | t1)
        if t2 & 0x80:
            temp1=temp1 | 0xFFFF0000
        return temp1/16
```

Notice that the function returns `-1000` if there is no device, `-2000` if there is a CRC error and `-3000` if the device fails to provide data. These values are outside the range of temperatures that can be measured.

The Complete Program

The complete program to read and display the temperature is:

```
from utime import sleep_ms, sleep_us
from machine import Pin
from time import sleep

class DS18B20:
    def __init__(self,pin):
        self.pin=Pin(pin,mode=Pin.IN)
        self.pin.high()

    def presence(self):
        self.pin.init(mode=Pin.OUT)
        self.pin.high()
        sleep_ms(1)
        self.pin.low()
        sleep_us(480)
        self.pin.init(mode=Pin.IN)
        sleep_us(70)
        b = self.pin.value()
        sleep_us(410)
        return b

    @micropython.native
    def writeBit(self,b):
        if b == 1:
            delay1 = 1
            delay2 = 30
        else:
            delay1 = 30
            delay2 = 0
        self.pin.low()
        for i in range(delay1):
            pass
        self.pin.high()
        for i in range(delay2):
            pass

    def writeByte(self,byte):
        self.pin.init(mode=Pin.OUT)
        for i in range(8):
            self.writeBit(byte & 1)
            byte = byte >> 1
        self.pin.init(mode=Pin.IN)
```

```
@micropython.native
def readBit(self):
    self.pin.init(mode=Pin.OUT)
    self.pin.low()
    self.pin.high()
    self.pin.init(mode=Pin.IN)
    b = self.pin.value()
    sleep_us(60)
    return b

def readByte(self):
    byte = 0
    for i in range(8):
        byte = byte | self.readBit() << i
    return byte

def convert(self):
    self.writeByte(0x44)
    for i in range(500):
        sleep_ms(10)
        if self.readBit() == 1:
            j=i
            break
    return j

def crc8(self,data,len):
    crc = 0
    for i in range(len):
        databyte = data[i]
        for j in range(8):
            temp = (crc ^ databyte) & 0x01
            crc >>= 1
            if temp:
                crc ^= 0x8C
            databyte >>= 1
    return crc
```

```
def getTemp(self):
        if self.presence()==1:
            return -1000
        self.writeByte(0xCC)
        if self.convert()==500:
            return -3000
        self.presence()
        self.writeByte( 0xCC)
        self.writeByte( 0xBE)
        data=[]
        for i in range(9):
            data.append(self.readByte())
        if self.crc8(data,9)!=0:
            return -2000
        t1 = data[0]
        t2 = data[1]
        temp1 = (t2 << 8 | t1)
        if t2 & 0x80:
            temp1=temp1 | 0xFFFF0000
        return temp1/16


dS18B20=DS18B20(2)
if dS18B20.presence() == 1:
    print("No device")
else:
    print("Device present")

print(dS18B20.getTemp())
```

A PIO DS18B20 Program

Implementing the 1-Wire protocol using a PIO is one of the most difficult tasks we tackle. It needs to both send and receive data and the master has to provide the start pulse for the slave to send data. The specification also makes it seem that we have to use a conditional to vary the timing of the data part of the pulse depending on sending a one or a zero. This is a lot of logic to fit into 32 instructions. In addition a 1-Wire device can send far more data than the FIFO can hold. At first appraisal the idea of implementing the 1-Wire protocol in a PIO program seems hopeless, but with a shift in emphasis it can be made easier. Given the fact that the big-banging interface to the 1-Wire bus is operating on the edge of what is possible, this makes it all the more worthwhile to implement a fast PIO-based program.

The main operations that you need to work with a 1-Wire device is to send a set of bytes and receive a set of bytes. In general, the timing requirements of the 1-Wire bus are at the bit level. That is, when you read data the device will generally store the data until you are ready for it. This means that if you can implement a byte read/write in the PIO then the main program can get on with something else while the 1-Wire device processes data. If the main program is interrupted then no harm is done as it can expect the PIO to continue to process the next byte ready for when it returns.

What all of this implies is that all we really have to do is implement a bit send and bit receive facility. Using the way that 1-Wire protocol is usually described, this would be very difficult. In other words, trying to implement the simple-minded bit-banging protocol as a PIO program isn't going to work. A slightly different interpretation of the protocol, however, makes it very possible. If you think of each bit frame as starting with a short low pulse which signals the start of the frame, then a zero sends a low after the initial pulse and a one sends a high. In other words, we can implement the write protocol as:

1) Send short "start bit" of about 6μs
2) Send the data bit for the rest of the frame – about 60μs
3) Set the line high for 10μs as a spacer between bit frames

The read protocol is much the same:

1) Send short "start bit" of about 6μs
2) After about 9μs sample the line and use this as the read bit
3) Do nothing for about 55μs as a spacer between bit frames

In this form the protocol doesn't need conditionals as the sending of a zero or a one and the receiving of a zero or a one follows the same steps.

Using this version you could simplify the logic of the bit-banging example given earlier, but there is no real need to as there is plenty of space and computational power – using a PIO is a very different matter and regularity in the implementation of a protocol is important.

There is also the problem of how to implement the PIO program so that the MicroPython program can make use of it. One possibility is to use one state machine for send and other to receive, but it is possible to create a single program to to do both jobs.

The PIO program starts off with a presence pulse and, as this is long compared to the clock time, we use a loop set by the MicroPython program. As you generally don't need a presence pulse when reading data from a device, we can use a zero value to indicate that the operation is a read, not a write. That is, you use the program by pushing data onto the FIFO – the first byte starts the program off and if it is zero it starts a read operation and otherwise starts a write operation with its value setting the length of the initialization pulse:

```
.program DS1820
.wrap_target
again:
       pull block
       mov x, osr
       jmp !x, read
write: set pindirs, 1
       set pins, 0
loop1:
       jmp x--,loop1
       set pindirs, 0 [31]
       wait 1 pin 0 [31]
```

Once the write part of the code gets started, the non-zero value passed in via the FIFO is used to create a long, 500μs, low pulse which the slave responds to with a 120μs low presence plus. The final instruction waits for the end of the pulse so that we can start to send some data.

The second byte pushed onto the FIFO gives the number of bytes to send and these are pushed onto the stack to follow. Notice that if there are more than four or five then the program might stall until the FIFO has space.

```
       pull block
       mov x, osr
bytes1:
       pull block
       set y, 7
       set pindirs, 1
bit1:
              set pins, 0 [1]
              out pins,1 [31]
              set pins, 1 [20]
              jmp y--,bit1
       jmp x--,bytes1

       set pindirs, 0 [31]
       jmp again
```

You can see that the inner loop starting at `bit1` takes the data byte in the OSR and sends it out to the GPIO line, a bit at a time. The first instruction in the loop generates the short start pulse and the second instruction sets the line high or low for 32 clock cycles. The final instruction in the loop provides the space between the bit frames. The loop repeats until all eight bits have been sent when the outer loop gets another byte to send, if there is one. If not, the program restarts and waits for the next set of data.

The read part of the program is very similar to the write:

```
read:
       pull block
       mov x, osr
bytes2:
       set y, 7
       bit2:
              set pindirs, 1
              set pins, 0 [1]
              set pindirs, 0 [5]
              in pins,1 [10]
       jmp y--,bit2
jmp x--,bytes2
.wrap
```

In this case the only data we need from the FIFO is the number of bytes to read. This is stored in the X register and controls the number of bytes processed. The inner loop, starting at `bit2`, reads each bit in turn. The first three instructions change the line to output, send the short start pulse and sets the line back to input. The `in` instruction samples the line after a short delay and gets the data which is auto-pushed onto the FIFO ready for the MicroPython program to read. Notice that if the MicroPython program fails to read the data then the loop stalls, which is fine as long as the 1-Wire device can live with a pause – the DS18B20 can.

The entire program is 29 instructions, which does leave space for instructions to toggle another GPIO line for debugging purposes. It is very probable that it can be made shorter by the use of setside operations.

Next we move on the MicroPython program. We first need an initialize things using the state machine constructor:

```
sm=rp2.StateMachine(0,DS1820,freq=490196, set_base=Pin(2),
                           out_base=Pin(2),in_base=Pin(2),
                            out_shiftdir=rp2.PIO.SHIFT_RIGHT,
                             in_shiftdir=rp2.PIO.SHIFT_RIGHT)
sm.active(1)
```

The frequency is selected to allow the pulse times to be implemented using delays from 1 to 31.

The PIO program is fairly general in that you could use it to talk to almost any 1-Wire device, but we are interested in the DS18B20 and it is easy enough to implement a function to read the temperature along the lines of the previous function:

```
def getTemperature(sm):
```

First we send an initialization pulse, followed by a Skip ROM and a Convert command. The first item sets a write operation with an initialization pulse:

```
sm.put(250)
sm.put(1)
sm.put(0xCC)
sm.put(0x44)
```

Next we should wait until the conversion is complete, but for simplicity we wait for 1s – after which time the conversion is either complete or the device is broken. Then we send another initialization pulse, a Skip ROM and a Read ScratchPad command:

```
sleep_ms(1000)
sm.put(250)
sm.put(1)
sm.put(0xCC)
sm.put(0xBE)
```

Finally we send a read command for nine bytes of data:

```
sm.put( 0)
sm.put( 8)
```

and read the nine bytes from the FIFO:

```
data=[]
for i in range(9):
    data.append(sm.get() >> 24)
return data
```

Now we have the same nine bytes of data as in the bit-banging example and the rest of the program is identical.

A better idea is to create two functions which read and write an array of bytes to the device:

```
def writeBytes(sm, bytes, len):
    sm.put(250)
    sm.put(len-1)
    for i in range(len):
         sm.put(bytes[i])

def readBytes(sm,len):
    sm.put(0)
    sm.put(len-1)
    bytes=[]
    for i in range(len):
        bytes.append(sm.get() >> 24)
    return bytes
```

Using these two functions the interaction with the 1-Wire device starts to look a lot like using the SPI or I2C bus. For example, the `readTemperature` function becomes:

```
def getTemperature(sm):
    writeBytes(sm,[0xCC,0x44],2)
    sleep_ms(1000)
    writeBytes(sm,[0xCC,0xBE],2)
    data=readBytes(sm,9)
    return data
```

As you can see this is now just a set of calls to general read/write functions. Notice that rather than test of the data to be ready we simply wait for 1000ms.

This is about as complex a PIO program as you can implement. This raises the question of what to do if you need something more complex? You might think that using additional state machines would help, but you are still limited to 32 instructions in total. There are two possible practical solutions. You can use immediately executed instructions put directly into the state machine from the MicroPython program. This works for small tasks such as initialization. The second solution is more generally useful. If you can split your program into small, less than 32 instructions, modules then you can load one into one PIO and another into the second PIO. This, of course, occupies both PIOs and you can't use a PIO for another task. For example, you could use one PIO to send data to a 1-Wire bus device and the other to receive data. This approach is limited to just two modules, so a total of 64 instructions. You can extend this idea to any number of modules if they can be reloaded into the same PIO. For example, by splitting the 1-Wire code into a receive and a send module you could easily afford the time to load each module and set it up into the same PIO.

The current program can only work with a single device on the 1-Wire bus. If you want to support more, you could implement the search algorithm outlined in Chapter 15 of *Raspberry Pi IOT in C, Second Edition,* though this is complex. A simpler solution is to use one state machine and one GPIO line per device. This can support up to eight different 1-Wire devices.

Complete Program

Of course, things are much easier to use when packaged as a class. The complete program listed below is a version of the earlier program that used bit-banging to implement a DS18B20 but now using the PIO program:

```
import rp2
from machine import Pin
from utime import sleep_ms

@rp2.asm_pio(set_init=rp2.PIO.OUT_HIGH, out_init=rp2.
                     PIO.OUT_HIGH, autopush=True,push_thresh=8)
def DS1820():
    wrap_target()
    label("again")
    pull(block)
    mov(x, osr)
    jmp(not_x, "read")
    label("write")
    set(pindirs, 1)
    set(pins,0)
    label("loop1")
    jmp(x_dec,"loop1")
    set(pindirs, 2) [31]
    wait(1, pin, 0) [31]

    pull(block)
    mov(x, osr)
    label("bytes1")
    pull(block)
    set( y, 7)
    set(pindirs, 3)
    label("bit1")
    set(pins, 0) [1]
    out(pins,1) [31]
    set(pins, 1) [20]
    jmp(y_dec,"bit1")
    jmp(x_dec,"bytes1")
    set(pindirs, 0) [31]
    jmp("again")
```

```
    label("read")
    pull(block)
    mov( x, osr)
    label("bytes2")
    set( y, 7)
    label("bit2")
    set(pindirs, 1)
    set(pins, 0) [1]
    set(pindirs, 0) [5]
    in_(pins,1) [10]
    jmp(y_dec,"bit2")
    jmp(x_dec,"bytes2")
    wrap()

class DS18B20:
    def __init__(self,pin):
        self.sm=rp2.StateMachine(0,DS1820,freq=490196,
                              set_base=Pin(2),out_base=Pin(2),
                 in_base=Pin(2),out_shiftdir=rp2.PIO.SHIFT_RIGHT,
                                  in_shiftdir=rp2.PIO.SHIFT_RIGHT)
        self.sm.active(1)

    def writeBytes(self, bytes, len):
        self.sm.put(250)
        self.sm.put(len-1)
        for i in range(len):
            self.sm.put(bytes[i])

    def readBytes(self,len):
        self.sm.put(0)
        self.sm.put(len-1)
        bytes=[]
        for i in range(len):
            bytes.append(self.sm.get() >> 24)
        return bytes

    def getTemp(self):
        self.writeBytes([0xCC,0x44],2)
        sleep_ms(1000)
        self.writeBytes([0xCC,0xBE],2)
        data=self.readBytes(9)
        t1 = data[0]
        t2 = data[1]
        temp1 = (t2 << 8 | t1)
        if t2 & 0x80:
            temp1=temp1 | 0xFFFF0000
        return temp1/16
dS18B20=DS18B20(2)
print(dS18B20.getTemp())
```

Other Commands

As well as the commands that we have used to read the temperature, the DS18B20 supports a range of other commands. There are two commands concerned with when there are more devices on the bus. Search ROM, `0xF0`, is used to scan the bus to discover what devices are connected and Match ROM, `0x55`, is used to select a particular device.

You can also read the unique 64-bit code of a device using the Read ROM command, `0x33`. In this case, the slave transmits eight bytes, comprised of a single-byte device family code, `0x28` for the DS18B20, six bytes of serial number and a single CRC byte.

As well as the Read ScratchPad command that we used to read the temperature, there is also a Write ScratchPad command, `0x4E`.

The format of the scratchpad is:

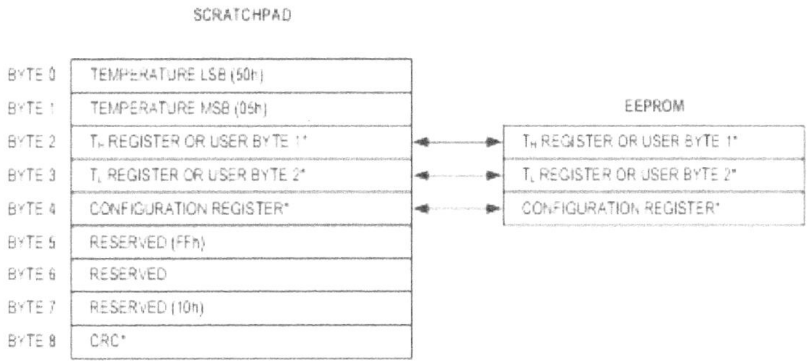

The first two bytes are the temperature that we have already used. The only writable entries are bytes 2, 3 and 4. The Write ScratchPad command transfers three bytes to these locations. Notice that there is no CRC and no error response if there is a transmission error. The datasheet suggests that you read the scratchpad after writing it to check that you have been successful in setting the three bytes.

The third byte written to the scratchpad is to the configuration register:

BIT 7	BIT 6	BIT 5	BIT 4	BIT 3	BIT 2	BIT 1	BIT 0
0	R1	R0	1	1	1	1	1

Essentially the only thing you can change is the resolution of the temperature measurement.

Configuration Register	Resolution	Time
0x1F	9 bits	93ms
0x3F	10 bits	175ms
0x5F	11 bits	375ms
0x7F	12 bits	750ms

The time quoted is the maximum for a conversion at the given precision. You can see that the only real advantage of decreasing precision is to make conversion faster. The default is `0x7F` and 12 bits of precision.

The first two bytes of the write scratchpad set a high and low temperature alarm. This feature isn't much used, but you can set two temperatures that will trigger the device into alarm mode. Notice you only set the top eight bits of the threshold temperatures. This is easy enough, but the alarm status is set with every read so if the temperature goes outside the set bounds and then back in the alarm is cleared.

The second problem is that, to discover which devices are in alarm mode, you have to use the Alarm Search command, `0xEC`. This works like the Search ROM command, but the only devices that respond are the ones with an alarm state. The alarm feature might be useful if you have a lot of devices and simply want to detect an out-of-band temperature. You could set up multiple devices with appropriate temperature limits and then simply repeatedly scan the bus for devices with alarms set.

You may notice that the scratchpad also has an EEPROM memory connected. You can transfer the three bytes of the scratchpad to the EEPROM using Copy Scratchpad, `0x48`, and transfer them back using the Recall EEPROM command, `0xB8`. You can use this to make the settings non-volatile.

Finally there is the Read Power Supply command, `0xB4`. If the master reads the bus after issuing this command a `0` indicates that there are parasitic powered devices on the bus. If there are such devices the master has to run the bus in such a way that they are powered correctly.

If you restrict yourself to a single slave device on the bus, this is more or less all there is to the DS18B20, and the 1-Wire bus in general. If you want to have multiple slave devices, however, while you don't need any more hardware, you do need some more software. Multidrop working as it is called is beyond the scope of this introduction, but you should find it possible to follow the description in Chapter 15 of *Raspberry Pi IOT in C, Second Edition* and implement your own version.

Summary

- The 1-Wire bus is a proprietary, but widely-used, bus. It is simple and very capable.
- As its name suggests it makes use of a single data wire and usually a power supply and ground.
- It is possible to dispense with the power line and the connected device will draw power from the data line.
- Implementing the 1-Wire protocol is mostly a matter of getting the timing right.
- There are three types of interaction: presence pulse, read and write.
- The presence pulse simply asks any connected devices to reply and make themselves known.
- The 1-Wire protocol is easier to implement than you might think because each bit is sent as a "slot" and while timing is critical within the slot, how fast slots are sent isn't and the master is in control of when this happens.
- The DS18B20 temperature sensor is one of the most commonly encountered 1-Wire bus devices. It is small, low-cost and you can use multiple devices on a single bus.
- After a convert command is sent to the device, it can take 750ms before a reading is ready.
- To test for data ready you have to poll on a single bit. Reading a zero means data not ready and reading a one means data ready.
- When the data is ready, you can read the scratchpad memory where the data is stored.
- The DS18B20 has other commands that can be used to set temperature alarms etc, but these are rarely used.
- With a shift in viewpoint, it is just possible to squeeze a 1-Wire bus protocol into a 32-instruction PIO program.

Chapter 16

The Serial Port

The serial port is one of the oldest of ways of connecting devices together, but it is still very useful as it provides a reasonably fast communication channel that can be used over a longer distance than most other connections such as USB. Today, however, its most common and important use is in making connections with small computers and simple peripherals. It can also be used as a custom signal decoder, see later.

Serial Protocol

The serial protocol is very simple. It has to be because it was invented in the days when it was generated using electromechanical components, motors and the like. It was invented to make early teletype machines work and hence you will often find abbreviations such as TTY used in connection with it. As the electronic device used to work with serial is called a Universal Asynchronous Receiver/Transmitter, the term UART is also often used.

The earliest standards are V24 and RS232. Notice, however, that early serial communications worked between plus and minus 24V and later ones ±12V. Today's serial communications work at logic, or TTL, levels of 0V to 5V or 0V to 3.3V. This voltage difference is a problem we will return to later. What matters is that, irrespective of the voltage, the protocol is always the same.

For the moment let's concentrate on the protocol. As already mentioned, the protocol is simple. The line rests high and represents a zero. When the device starts to transmit it first pulls the line low to generate a start bit. The width of this start bit sets the transmission speed - all bits are the same width as the start bit. After the start bit there are a variable number, usually seven or eight, data bits, an optional single parity bit, and finally one or two stop bits.

Originally the purpose of the start bit was to allow the motors etc to get started and allow the receiving end to perform any timing corrections. The stop bits were similarly there to give time for the motors to come back to their rest position. In the early days the protocol was used at very slow speeds; 300 baud, i.e. roughly 300 bits per second, was considered fast enough.

Today the protocol is much the same, but there is little need for multiple stop bits and communication is often so reliable that parity bits are dispensed with. Transmission speeds are also higher, typically 9600 or 115200 baud.

To specify what exact protocol is in use, you will often encounter a short form notation. For example, 9600 baud, 8 data bits, no parity, one stop bit, will be written as 9600 8n1. Here you can see the letter 0 (`01101111` or `0x6F`) transmitted using 8n1:

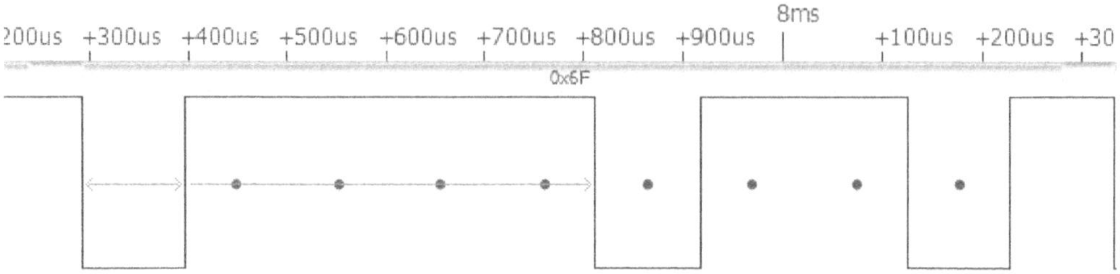

Notice that the signal is sent least significant bit first. The first low is the start bit, then the eight dots show the ideal sampling positions for the receiver. The basic decoding algorithm for receiving serial data is to detect the start of the start bit and then sample the line at the center of each bit time. Notice that the final high on the right is the stop bit.

Notice also that the sampling points can be put to use on custom protocols. As long as the data is transmitted in fixed time "cells" indicated by a start bit, you can use a serial port to read individual bits.

For a serial connection to work, it is essential that the transmitter and the receiver are set to the same speed, data bits and parity. Serial interfaces most often fail because they are not working with the same settings.

A logic analyzer with a serial decoder option is an essential piece of equipment if you are going to do anything complicated with serial interfacing.

What is a baud? Baud rate refers to the basic bit time. That is, 300 baud has a start bit that is 1/300s wide. This means that for 9600 baud a bit is 1/9600 wide or roughly 104μs and at 115200 baud a bit is 1/115200 or roughly 8.6μs. Notice that baud rate doesn't equate to speed of sending data because of the overhead in stop, start and perhaps parity bits to include in the calculation.

UART Hardware

A simple serial interface has a transmit pin, TX, and a receive pin, RX. That is, a full serial interface uses two wires for two-way communications. Typically you connect the TX pin on one device to the RX pin on the other and vice versa. The only problem is that some manufacturers label the pins by what they should be connected to not what they are and then you have to connect RX to RX and TX to TX. You generally need to check with a scope, logic probe or meter which pin is which if you are in any doubt.

In addition to the TX and RX pins, a full serial interface also has a lot of control lines. Most of these were designed to help with old fashioned teleprinters and they are not often used. For example, RTS - Request To Send is a line that it used to ask permission to send data from the device at the other end of the connection, CTS - Clear To Send is a line that indicates that it is okay to send data and so on. Usually these are set by the hardware automatically when the receive buffer is full or empty.

You can use RTS and CTS as a hardware flow control. There is also a standard software flow control involving sending XON and XOFF characters to start and stop the transmission of data. For most connections between modern devices you can ignore these additional lines and just send and receive data. If you need to interface to something that implements a complex serial interface you are going to have to look up the details and hand-craft a program to interact with it.

The Pico has two UARTs which can be used in addition to the USB interface. Both UARTs support buffered (32-character) and unbuffered modes and RTS/CTS hardware flow control. The UARTs can be connected to a group of four GPIO pins:

UART0				
TX	GP0	GP12	GP16	GP28
RX	GP1	GP13	GP17	GP29
CTS	GP2	GP14	GP18	
RTS	GP3	GP15	GP19	

UART1				
TX	GP4	GP8	GP20	GP24
RX	GP5	GP9	GP21	GP25
CTS	GP6	GP10	GP22	GP26
RTS	GP7	GP11	GP23	GP27

You can select any combination of pins and if you don't want to use the RTS/CTS lines they can be used for some other purpose. At the moment MicroPython doesn't support the use of RTS/CTS.

There is a problem with making the connection to the Pico's RX and TX pins in that devices work at different voltages. PC-based serial ports usually use +13V to -13V and all RS232-compliant ports use a + to - voltage swing, which is completely incompatible with most microprocessors which work at 5V or 3.3V.

If you want to connect the Pico to a PC or other standard device then you need to use a TTL-to-RS232 level converter. In this case it is easier to use the PC's USB port as a serial interface with a USB-to-TTL level converter. All you have to do is plug the USB port into the PC, install a driver and start to transmit data.

Remember when you connect the Pico to another device that TX goes to the other device's RX and RX goes to the other device's TX pin. Also remember that the signaling voltage is 0V to 3.3V.

Setting Up the UART

To create a `UART` object you use its constructor:

```
uart = UART(id=, baudrate=, bits=, parity=, stop=, tx=, rx=)
```

and to stop using the `UART`:

```
uart.deinit()
```

The `tx` and `rx` pins are specified as Pin objects and have to be correct according to the table given earlier.

You can set 5, 6, 7 or 8 data bits and 1 or 2 stop bits. Parity can be any of:

`None` = none

`0` = even

`1` = odd

By default 32-element FIFO buffers are used on RX and TX.

There are many MicroPython features not currently supported. You cannot use the `init` method, no access to CTS/RTS, no support for timeout and only very simple buffer access.

Data Transfer

After you have set up the UART you can start sending and receiving data using methods. There are three general-purpose read functions:

```
read(nbytes)
readinto(buf,nbytes)
readline()
```

In the first two methods data is stored in a byte array. In `readline` bytes are read in until a newline character is encountered and a string is returned.

The `readinto` method is slightly more flexible. If you don't specify the number of bytes to read then the number read is determined by the size of the buffer, i.e. a buffer's worth of data is read in. If you specify the number of bytes less than the size of the buffer then only that number of bytes are read. The function also returns the number of bytes read.

There is a single write method:

```
write(buf)
```

this writes the number of bytes in the bytearray `buf` and returns the number of bytes written.

These methods are more difficult to use than you might expect. The problem is that they don't allow a timeout to be specified and this is often essential in a UART-based protocol. In addition, the functions work with the FIFO and will block if there is no data to read or space to write in the FIFO buffer. If there is no external device to send or read data, this means that any of the read or write functions could hang the machine indefinitely.

Also notice that in practice the `nbytes` parameter in `read` and `readinto` is optional, but if you don't specify the number of bytes then the method will hang forever trying to gather data.

If you connect GP4 to GP5 we can use these pins to perform some loopback tests that demonstrate some of the problems:

```
from machine import UART,Pin
from utime import sleep_ms

uart=UART(1,baudrate=9600,bits=8,parity=2,rx=Pin(5),tx=Pin(4))
SendData = bytearray("Hello World \n","utf-8")
uart.write(SendData)
RecData=uart.read()
print(RecData)
```

Notice that we are using UART1 because UART0 is the alternative REPL port for MicroPython. However it isn't used very often so feel free to use it if UART1 isn't suitable for some reason.

The write function sends 13 bytes of data which are received and stored in the FIFO buffer ready for the read function to retrieve. As there is no feedback from the receiver, the transmitter carries on sending data out even if the receiver's buffer is full and the send bytes are just being ignored. After all of the data has been sent, the read method reads in as much data as it needs to, but the receive buffer may well be full and data may have been lost.

As it is the program stalls anyway. The `read()` function attempts to read as many items of data as it can and so it never returns. You can make the program work by specifying the number of characters you want to read. If you change the function to:

```
RecData=uart.read(13)
```

you will now see the full message printed. The reason this works is that the sent data is 13 bytes long and this is small enough to be stored in the receive buffer.

If you try:

```
from machine import UART,Pin
from utime import sleep_ms

uart=UART(1,baudrate=9600,bits=8,parity=2,rx=Pin(5),tx=Pin(4))
SendData = bytearray("A"*32,"utf-8")
uart.write(SendData)
sleep_ms(500)
RecData=uart.read(32)
print(RecData)
```

you will find that everything works, but if you change `32` to `33` the program hangs. The reason is that the receive buffer can hold 32 bytes. If you try to send 33 bytes or more then all of the bytes after the first 32 are lost and when the `read` tries to read 33 bytes it waits forever because there are only 32 bytes to read in the buffer.

Also notice that the read operation will not begin until the write method returns and this only happens after it has sent the full 66 bytes. There is no way to engineer the `read` to clear some of the RX FIFO before it fills in this simple approach.

You might think that a rule like “never read more than 32 bytes” would be useful, but this just covers the case of receive buffer overflow. You can never be certain that any number of bytes will be available to read because you are relying on outside sources. A better, though not perfect rule, is never send more than 32 bytes – as this at least means that if the RX FIFO starts empty no bytes will be lost.

As an illustration of how complicated things can get try the same program with `sleep_ms` removed. In this case you will find that it works up to 65 but hangs at 66 – it looks as if the buffer is 65 characters in size not 32. What is happening? The answer is that, without the sleep, the write command finishes after sending all 65 characters and then the read command takes over. The read buffer has already stored the first 32 bytes of the transmission in the RX FIFO which the read command starts to read, but while this is happening the TX FIFO buffer is still sending another 32 bytes – serial transmission at 9600 baud is slow – which there is now space in the RX buffer for so the read manages to get in 65 characters not just 32. If you go beyond 65 characters then then some characters are lost before the read command gets a chance to start emptying the RX FIFO.

The problem of stalling when trying to read or write a serial interface is serious as you can never know if the outside world is going to behave as you expect. The usual approaches to tackling the problem are polling or using interrupts. As the Pico's version of MicroPython doesn't implement UART interrupts, this leaves polling.

Serial Polling

The key method in implementing UART polling is:

```
any()
```

which returns 0 if there are no characters in the RX buffer and 1 if there are. Notice it doesn't return the number of characters that are waiting in the buffer.

Unfortunately, there is no equivalent method to test to see if there is space in the TX Buffer, but it is easy to add one – see later.

It is easy to see that you can use this function to write polling loops which test to see if there is anything to read. For example:

```
def uart_read(uart):
    rxData = bytes()
    while uart.any() > 0:
        rxData += uart.read(1)
    return rxData
```

The advantage of this loop is that you can do other work within the `while` loop and check that reading hasn't been waiting too long. That, is you can easily extend this to include a timeout. However there are problems with this function.

For example:

```
from machine import UART,Pin,mem32
from utime import sleep_ms

def uart_read(uart):
    rxData = bytes()
    while uart.any() > 0:
        rxData += uart.read(1)
    return rxData

uart=UART(1,baudrate=9600,bits=8,parity=2,rx=Pin(5),tx=Pin(4))
SendData = bytearray("A"*65,"utf-8")
uart.write(SendData)
sleep_ms(500)
RecData=uart_read(uart)
print(RecData)
print(len(RecData))
```

This program sends 65 "A"s of which all but 32 are lost and then `uart_read` reads in as many characters as are available in the FIFO, i.e. 32 characters. The advantage of this is that the function only reads until there is no more data to read – the problem is that the external transmitter could send continue to send data while `uart_read` is reading the RX FIFO and the function would read these as well as what was already in the RX FIFO. This is the behavior you want, but what if the transmitter pauses for long enough for the reading function to empty the RX FIFO? In this case the reading function would assume that the transmission was over and stop reading data even if the transmitter started sending again – this might not be what we want.

Timeouts

From the previous section you should be able to see that there the usual way that a transmitter and receiver interact is that the transmitter, after perhaps a longish silence "decides" to send some data. This is usually sent as a block and, as long as the transmitter isn't overloaded, each character follows the next with minimum delay. What this means is that, from the receivers point of view there might be a long wait before being able to start reading data, but after that there should be only a short interval between each character. If the receiver finds itself waiting for a long time for the next character then the chances are the transmitter has finished sending a block of data.

Thus there are two sorts of "timeout" we need to specify. An initial timeout, which is how long to wait before the first character is received, and an inter-character timeout ,which is the maximum time the receiver should wait between characters before concluding that in all probability the transmitter has finished. It is fairly easy to change the `uart_read` function to include an initial timeout and an inter-character timeout:

```
def uart_read(uart,timeout,interchar):
    rxData = bytes()
    t=ticks_us()
    while True:
        if uart.any()>0:
            break
        if ticks_us()-t>timeout:
            return None
    while True:
        while uart.any() > 0:
            rxData += uart.read(1)
            t=ticks_us()
        if ticks_us()-t>interchar:
            break
    return rxData
```

For example:

```
RecData=uart_read(uart,1000,100)
print(RecData)
```

will wait 1ms for the first character to arrive and then 0.1ms for each subsequent character to arrive. The time to wait for the first character is usually set by a combination of how much time the receiver can devote to waiting in a polling loop and how often the transmitter sends data. The inter-character time is usually low if the transmitter manages to keep the RX FIFO topped up. It should be set to less than the characteristic time between data blocks from the transmitter.

Polling On Write

Writing data to the UART is much simpler than dealing with reading it. Data is written to the TX FIFO as fast as it can be and the only thing that can go wrong is having to wait because the FIFO is full. Give that the FIFO is 32 characters this is likely to happen whenever you are transmitting a large block of data. If your program has nothing else to do while the transmission is underway, there is no problem. If, however, your program has to service a polling loop doing a range of other things then it can be a big problem.

The solution is to test the status of the TX FIFO. As already mentioned MicroPython doesn't have a method to do this job, but we can write one to read the appropriate status bit in the UART's register:

```
from machine import UART,Pin,mem32
def uart_is_writeable(uart):
    if uart==0:
        base=0x40034000
    elif uart==1:
        base= 0x40038000
    else:
        return None
    return not((mem32[base+0x018] & 0x00000020))
```

Using this function we can write a non-blocking write:

```
def uart_write_noblock(uart,uartNo,char):
    if uart_is_writeable(uartNo):
        uart.write(bytes([char]))
        return True
    return False
```

This is slightly more awkward than it would ideally be due to the need to pass the UART object and its index number – there seems to be no easy way to discover what index number was used to create a UART object. The `char` parameter also has to be converted to a `bytes` object to be passed to the write function. This function returns `True` if it successfully sent the character and `False` if the TX FIFO is full. If you try this out using:

```
SendData = bytearray("A"*65,"utf-8")
for i in range(len(SendData)):
    uart_write_noblock(uart,1,SendData[i])
sleep_ms(500)
RecData=uart_read(uart,1000,100)
print(RecData)
```

you will discover that only 32 characters are sent after the TX FIFO is full. The important point is that the `uart_write_noblock` function always returns immediately and this means you can do work within the loop while waiting for space in the TX FIFO buffer to become available.

Reading and Writing Characters and Strings

In the previous examples we have used a bytearray to read and write data. A more basic method of working is to read and write a single character at a time. While there are no easy-to-use UART methods to read and write a character, the functions:

```
def uart_putc(uart,char):
    uart.write(char[0])

def uart_getc(uart):
    char=uart.read(1)
    return char.decode("utf-8")
```

do the job. They are both blocking, but it is easy to convert them to non-blocking using the techniques in the previous section.

You can use the provided UART methods to send and receive simple strings. For example:

```
uart.write("Hello World \n")
print(uart.readline())
```

Notice that we need the newline character at the end of the string so that the `readline` method will return.

Finally, it is worth mentioning that there is a function that will send a break signal. The break signal doesn't correspond to any character that you can send in the usual way. Instead it corresponds to holding the TX line low for more than the time to transmit a single character. It is not often used today, but there are serial-based protocols that use the break to synchronize data transfer. You can send such a break using:

```
sendbreak()
```

Summary

- The serial port is one of the oldest ways of connecting devices together, but it is still very much in use.
- The serial protocol is asynchronous, but simple. A start bit gives the timing for the entire exchange.
- There are many control lines once used with telephone equipment that are mostly ignored in computer use. Similarly, the original ±12V signaling has been mostly replaced by 5V, and even 3.3V, signaling.
- The standard hardware that implements a serial connection is usually called a UART.
- The Pico contains two UARTs in addition to the USB serial connection.
- MicroPython provides functions for initializing and sending and receiving data.
- Each UART has a 32-element FIFO send and receive buffer which can be disabled.
- You can use the read and write blocking functions to send and receive byte buffers of data.
- The built-in methods are all blocking and while they don't have facilities for timeouts, you can construct functions to implement them using the basic methods.
- With the help of a custom function you can also test to see if the TX FIFO has space and so create a non-blocking write functionnd also create functions to work with single characters and strings.
- As the UART buffers are small, all transactions should be in terms of amounts of data that do not fill the buffers.

Chapter 17

WiFi Using The ESP8266

The Pico currently lacks WiFi which severely limits its utility as an IoT device. The ESP8266 provides a low cost and relatively easy way to remedy this. As we will make a lot of use of the Pico's serial ports, it is assumed that you know roughly how these work. In particular, it is assumed that you are familiar with the material in the previous chapter.

The Amazing ESP8266

Before we get on to the details of using the ESP8266 we need to find out what makes it special. It is remarkably low cost, $5 or less, but it is a full microprocessor with GPIO lines, RAM and built-in WiFi. It is built by a Chinese company Espressif Systems, but there are a number of copies on the market. The proliferation of devices and software revisions makes it difficult to work with, but it is well worth the effort.

While you can set up a development system yourself and program the ESP8266 to do almost anything, it comes with built-in software that allows it to be used as a WiFi module for other processors. This is how we are going to use it to give WiFi capability to the Pico.

The ESP8266 connects to the outside world using the serial port. The Pico controls it and transfers data using a system of AT commands. These were commonly used to control modems and other communication equipment and they still are used in mobile phone modems. In essence the ESP8266 looks like a modem that connects to WiFi. The module that is used in this chapter is the ESP-01, which is widely available from many different sources, but they all look like this:

There is another version, the ESP-07, which comes with a screen RF stage and other advantages, and this should also work. A bigger problem is that there are new versions of the firmware and some of these might not work in exactly the same way. However, it should be easy to make the changes necessary.

Connecting the ESP8266 ESP-01

There a number of minor problems in using the ESP8266. The first is that it comes with an 8-pin male connector which is not prototype board friendly. The best solution to this is to use some female-to-male jumper cables to connect it to the prototype board or use female-to-female cables to connect directly to the Pico. Note that for high baud rates cables should be kept as short as possible.

You can power the ESP8266 directly from the Pico's 3.3V supply pin, but if possible it is better to use an alternative as, when transmitting, the ESP8266 takes a lot of current, 300mA or so, and this means there isn't a lot left over to power other things. The maximum current for the Pico's 3.3V isn't given, but the chip used claims to be happy at 1200mA and it worked to run all of the examples in this chapter.

The pinout of the ESP-01 is usually shown from the component side, but in fact the pins that you want to connect to are on the other side. To make things easier, the two views are given in the diagram:

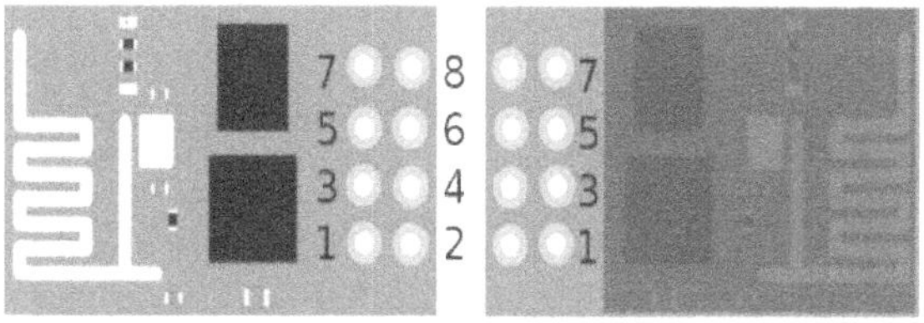

The pin functions are:

1 Ground	Connect to ground
2 TX	Serial Tx pin
3 GPIO2	Ignore
4 CHPD	Chip enable, connect to 3.3V
5 GPIO0	Ignore
6 RST	Reset, leave unconnected
7 RX	Serial Rx pin
8 VDD	Supply voltage, connect to 3.3V

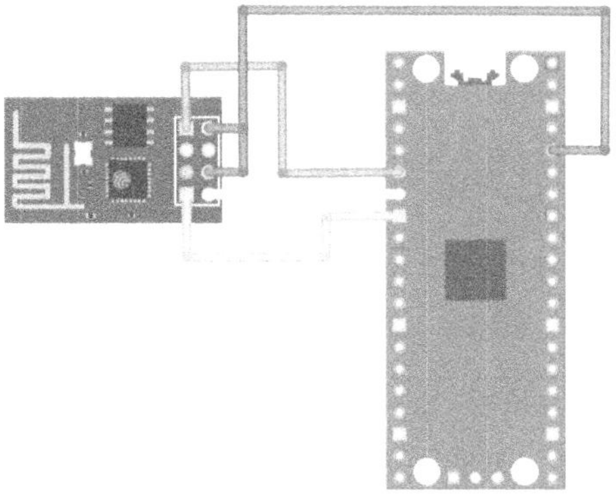

From the pinouts you should be able to work out the way the ESP8266 has to be connected. If we use GP4 as TX from the Pico and GP5 as RX to the Pico, we have to connect ESP-01 pin 7 to GP4, pin 2 to GP5 and pins 8 and 4 to the power supply. To make it all work we also have to connect pin 1 to the Ground.

AT Commands

The key idea in using the ESP8266 is that the Pico has to send an AT command, literally the characters AT, followed by other command strings. The command has to end with `\r\n` for the ESP8266 to take notice of it.

You can find a fill list of commands at the Espressif web site, but the most important are:

`AT`	Attention
`AT+RST`	Reset the board
`AT+GMR`	Firmware version
`AT+CWMODE=`	Operating Mode `1` - Client, `2` -Access Point, `3` - Client and Access Point
`AT+CWJAP=`	Join network
`AT+CWLAP`	View available networks
`AT+CWQAP`	Disconnect from network
`AT+CIPSTATUS`	Show current status as socket client or server
`AT+CIPSTART=`	Connect to socket server
`AT+CIPCLOSE`	Close socket connection
`AT+CIFSR`	Show assigned IP address when connected to network
`AT+CIPMUX=`	Set connection `0` - Single Connection, `1` - Multi-Channel Connection
`AT+CIPSERVER=`	Open the Socket Server
`AT+CIPMODE=`	Set transparent mode
`AT+CIPSTO=`	Set auto socket client disconnect timeout from 1s to 28800s
`+IPD`	Data

This is just a very general overview and omits the commands that allow the device to work as an access point. It is assumed that client mode is the more common application, but it isn't difficult to extend this example to access point operation.

Setup

We need a function to set up the hardware ready to communicate with the device:

```
def initWiFi(uartNo):
    uart=UART(uartNo,baudrate=115200,bits=8,stop=1,
                              parity=None,rx=Pin(5),tx=Pin(4))
    sleep_ms(100)
    return uart
```

The model of ESP8266 used worked at 115200 baud by default. Newer models and updated firmware are reported to work at 9600 baud. If this is the case you need to modify the data rate and/or change the ESP8266's baud rate.

The 100ms delay at the end is needed to allow the ESP8266 time to initialize.

Attention!

To start with the easiest, but possibly least useful, command, let's implement some functions that test that we have a connection to the device - always a good idea at the beginning of a project.

In this section we develop the first of a number of functions that send a command to the device and get back a response. To do this we need a few other functions to help and these are reused in other AT functions. Once you have one AT function, the rest are very similar, so it is worth spending some time following how this most simple AT function works.

There isn't much point in moving on until you have tested that the connection works. To do this send the string `AT\r\n` and the ESP8266 should reply with a single "OK", which proves that the serial interface is working.

The most basic AT function is simply:

```
def ATWiFi(uart):
       SendData= "AT\r\n"
       uart.write(SendData)
```

while this is the most basic form of function that will do the job, it isn't really practical. We need to check that it works and get some feedback by reading the data from the device.

We need the timeout read function given in the previous chapter, but modified to return a string:

```
def uart_read(uart,timeout,interchar):
    rxData = bytearray(2000)
    t=ticks_us()
    while True:
        if uart.any()>0:
            break
        if ticks_us()-t>timeout:
            return None
    while True:
        while uart.any() > 0:
            rxData += uart.read(1)
            t=ticks_us()
        if ticks_us()-t>interchar:
            break
    return rxData.decode("utf-8")
```

In this case an inter-character timeout of 0.1s is enough for more than 500 characters at 115200 baud and 100 at 9600 baud. The function attempts to read as much data as possible until an inter-character timeout occurs. We also have to change the way that the `rxData` structure is allocated. In the previous chapter we used a `bytes` object which changed its size as required. At 115200 baud the dynamic resizing of the bytes object is too slow and the result is that input characters are missed. The simplest solution is to switch to a preallocated bytearray of 2000 characters. This can be used as if it was dynamic, but it is much faster as memory doesn't have to be allocated up to the limit of 2000 characters.

If the AT command has been successful it returns a string containing `OK`. One of the problems of working with AT commands is that they don't have an exact syntax and there is no formal "end of message" signal. By observation, any successful command does end with `OK` and if you can't find "OK" in the response then you either haven't read it all or there has been an error.

With all of this we can now create an `ATWiFi` function that returns its response:

```
def ATWiFi(uart):
    SendData= "AT\r\n"
    uart.write(SendData)
    return uart_read(uart,10000,100)
```

A complete test program is:

```
from utime import ticks_us, sleep_ms
from machine import UART, Pin

def uart_read(uart, timeout, interchar):
    rxData = bytes()
    t = ticks_us()
    while True:
        if uart.any() > 0:
            break
        if ticks_us()-t > timeout:
            return None
    while True:
        while uart.any() > 0:
            rxData += uart.read(1)
            t = ticks_us()
        if ticks_us()-t > interchar:
            break
    return rxData.decode("utf-8")

def initWiFi(uartNo):
    uart = UART(uartNo, baudrate=115200, bits=8, stop=1,
                parity=None, rx=Pin(5), tx=Pin(4))
    sleep_ms(100)
    return uart

def ATWiFi(uart):
    SendData = "AT\r\n"
    uart.write(SendData)
    return uart_read(uart, 10000, 100)

uart = initWiFi(1)
response = ATWiFi(uart)
print(response)
```

When you run this program you should see:

```
AT
OK
```

printed on the serial console. If you don't there are five possible reasons:

1. You have connected the wrong pins - check
2. You haven't connected enable to +V
3. The power supply you are using is inadequate - check/replace
4. The serial console isn't working - check you can see any message
5. The baud rate is wrong - try 9600
6. The ESP8266 is broken - try another

Some Utility Functions

Assuming that you have managed to make the AT command work, it is time to move on to other AT commands. At this point it is also worth creating a WiFi object incorporating the functions we have so far:

```
from utime import ticks_us, sleep_ms
from machine import UART, Pin

class WiFi:
    def __init__(self,uartNo,tx,rx):
        self.uart = UART(uartNo, baudrate=115200, bits=8, stop=1,
                        parity=None, rx=Pin(rx), tx=Pin(tx))
    sleep_ms(100)

    def _read(self,timeout, interchar):
        rxData = bytes()
        t = ticks_us()
        while True:
            if self.uart.any() > 0:
                break
            if ticks_us()-t > timeout:
                return None
        while True:
            while self.uart.any() > 0:
                rxData += self.uart.read(1)
                t = ticks_us()
            if ticks_us()-t > interchar:
                break
        return rxData.decode("utf-8")

    def AT(self):
        self.uart.write("AT\r\n")
        return self._read(10000, 100)

wifi=WiFi(1,4,5)
response = wifi.AT()
print(response)
```

Notice that this form of the class isn't necessarily the best for a given purpose, but in this case it is best for understanding what is going on.

The first useful command is to find the Version number of the firmware. This is more or less the same as the AT function, but the command is `AT+GMR`:

```
    def getVersion(self):
        self.uart.write("AT+GMR\r\n")
        return self._read(10000, 100)
```

The device in use returned:

```
AT+GMR
AT version:1.3.0.0(Jul 14 2016 18:54:01)
SDK version:2.0.0(5a875ba)
v1.0.0.3
Mar 13 2018 09:35:47
OK
```

The manual says that it should return AT, SDK and the time it was compiled, so we get a little extra with this third-party device.

Reset is another useful function. The device often gets stuck and then a reset is all that you can try. This is a software reset; another option is to use a GPIO line to connect to reset pin 6. By controlling this line you can force the device to hard reset. The soft reset command is `AT+RST` and all you get back in theory is "OK", but in practice what the device sends varies:

```
def reset(self):
     self.uart.write("AT+RST\r\n")
     return self._read(10000, 100)
```

The final utility function is to set the serial connection parameters to 115200, 8 bits, 1 stop bit, no parity, no flow control:

```
def setUART(self):
    self.uart.write("AT+UART_CUR=115200,8,1,0,0\r\n")
    return self._read(10000, 100)
```

If you change the baud rate to something other than what is in use you will, of course, lose communication with the device until you reconfigure the serial connection.

You may notice that these utilities have a standard form – send a string, receive the response, and you can factor this out into another utility function if you want to, but the gains are small.

Configuring WiFi

The first thing we need to configure is the operating mode. The ESP8266 can operate as an access point, i.e. it can allow other devices to connect to it. However, in most cases you will want it to work in client mode, i.e. connecting to your existing WiFi.

Mode

A function to set its operating mode is:

```
def mode(self, mode):
    command="AT+CWMODE_CUR="+str(mode)+"\r\n"
    self.uart.write(command)
    return self._read(10000, 100)
```

In this case the command is:

```
AT+CWMODE_CUR = n
```

where `n` is `1` for client, `2` for access point and `3` for both. If you want to make the change permanent then change `CUR` to `DEF` and the setting is stored in flash memory. Older devices do not support `CWMODE_CUR`. Simply change it to `CWMODE`, which is deprecated.

To set client you would call:

```
modeWiFi(buf, 512, 1)
```

and see `OK` sent back.

Scan

The scan function is one that everyone wants to try out, but in practice it isn't very useful. Why would you want a list of WiFi networks? There are some applications for this function, but not as many as you might think. In most cases, you simply want to connect to a known WiFi network, which is what we do in the next section.

The scan command is easy - just send `AT+CWLAP` and the device sends you a complete list of WiFi networks. The problem is that scanning takes a long time and often hangs for a few seconds. The data is sent to the Pico in a series of blocks not one big block. How to deal with this depends on what you need. The simplest solution is to set the inter-character timeout to be larger than the typical time between blocks – 20s say. This will read all of the blocks and return them as a single string. The disadvantage of this method is that the return data could be large and you cannot process each block of information until all blocks are complete. For example, if you want to show the user the available WiFi networks then this approach will show nothing at all for a long time and then show all of the found WiFi networks in one go.

We can implement the scan function using a long inter-character timeout of 20s, but notice that nothing will be returned until all blocks are received:

```
def scan(self):
   self.uart.write("AT+CWLAP\r\n")
   return self._read(100000, 1000*1000*20)
```

If you want to process blocks as they arrive then you need to set the inter-character timeout to a realistic time between the characters in a block and set the overall timeout to be bigger than the time between blocks. In this scheme of things the read function will return after a single block and it is up to you to process that block quickly and then restart the read for the next block. The task is complete when the read function returns with an initial timeout. An alternative version of the method making use of reading multiple blocks is:

```
def scandisplay(self):
    self.uart.write("AT+CWLAP\r\n")
    block=self._read(10000, 100)
    print(block)
    for i in range(20):
        block=self._read(1000*1000*20, 100)
        if block==None:
            break
        print(block)
```

In this case the function is set to retrieve a maximum of 20 blocks and waiting up to 20s for each block, which could mean waiting as long as 400 seconds for the function to exit. However, in most cases it exits after only a small number of blocks. If you want to terminate the function sooner check for the “OK” response in each block and break out of the loop if found.

Connecting to WiFi

Our final, and most useful, functions connect the device to a known WiFi network. All you have to do is supply the SSID and password. There are other versions of the command that allow you to specify the connection more precisely, but this general form is the most useful. Connection to a network takes a while and there is quite a lot of data sent back, so we need to use a long inter-character timeout to make sure that we display all of the blocks:

```
def connect(self,ssid,password):
    command="AT+CWJAP_CUR=" +'"'+ ssid+'"' +","+'"'+
                                    password+'"'+ "\r\n"
    self.uart.write(command)
    for i in range(20):
        block=self._read(1000*1000*20, 100)
        if block==None:
            break
        if block.find("OK")>=0:
            break
    print("connected")
```

The only difficult part of the method is the use of different quote symbols to encode the quotes into the command string. If you have an older device you might need to change `CWJAP_CUR` to the deprecated `CWJAP` command. There is also a `CWJAP_DEF` command that will save the connection in the flash memory.

The connection is made with:

```
wifi.connect(buf, 512,"myWiFi","myPassword")
```

After a few seconds you should see:

```
AT+CWJAP_CUR="myWiFi","myPassword"
WIFI CONNECTED
WIFI GOT IP
OK
```

Once you are connected and the "WIFI GOT IP" message has been received you can ask what the IP address is:

```
def getIP(self):
    self.uart.write("AT+CIFSR\r\n")
    return self._read(100000, 100)
```

Of course, if you really need to know the IP address within a program you need to extract it from the string. The device replies with:

```
AT+CIFSR
+CIFSR:STAIP,"192.168.253.4"
+CIFSR:STAMAC,"5c:cf:7f:16:97:ab"
OK
```

This makes it very easy to get the IP address, even without the help of a regular expression.

Getting A Web Page

Now we have enough functions to tackle the two standard tasks in using the TCP stack, getting and sending data as a client and as a server.

First we tackle the problem of acting as a client. This isn't as common a requirement as you might expect because most of the time devices like the Pico are used to supply data to other servers, not the other way round. However, it is worth seeing how it is done.

It doesn't matter if you are implementing a client or a server, either way you make use of sockets which represent the basic TCP connection. What you do with this connection is up to you. For example, if you send HTTP headers on an appropriate port, you can fetch or deliver a web page, i.e. HTTP over TCP.

The first thing we have to do is set up a socket connection between the client, i.e. the Pico, and the server.

```
def getWebPage(self,URL,page):
    command= 'AT+CIPSTART="TCP",'+'"'+URL+'"'+",80\r\n"
    self.uart.write(command)
    block=self._read(1000*1000*2,1000*1000)
    if block.find("OK")<0:
        return -1
```

You pass the URL to the function as an IP address or as a full URL, but the device looks up domain names using a fixed set of DNS servers. It is recommended that you use an IP address, especially when testing. The `CIPSTART` command opens a socket to the specified IP address and port.

You can also specify a TCP or UDP connection in the AT command:

```
AT+CIPSTART=type, IP, port
```

In this case we open port 80 on the specified IP address or URL. If it works you will get back a message something like:

```
Connect
AT+CIPSTART="TCP","192.168.253.23",80
CONNECT
OK
```

Now we have a socket open we can send some data to the server and wait for some data to be sent back to us. This can be a problem as you can't anticipate the amount of data you get back from the web server.

For this example, the web page is served by a small sensor that returns a JSON temperature and Humidity reading. The sensor is another Pico running the web server as described in the next section.

To send data over a socket you use `CIPSEND`, which will send any data you specify to the server. As already made clear, what you send is a matter of what protocol you are using with the socket. In this case it is HTTP and we are going to send headers corresponding to a `GET` request for `index.html`

```
command='GET ' + page + ' HTTP/1.1\r\nHost:'+URL+'\r\n\r\n'
```

This time we can't just send the string because we need to include the length of the string in the header we send to the server.

The headers we are sending are:

```
GET /index.html HTTP/1.1
Host:192.168.253.23
```

Remember, an HTTP request always ends with a blank line.

To send this request we use the `CIPSEND` command which specifies the number of characters that are to follow:

```
command2="AT+CIPSEND="+str(len(command))+ "\r\n"
        self.uart.write(command2)
        block=self._read(1000*1000*8,1000*1000*1)
```

Now we have to send the number of bytes/characters that we specified in the `CIPSEND`. but first we wait for a ">" to indicate that the device is ready to receive the data:

```
        if block.find(">")<0:
                return -1
        self.uart.write(command)
        block=self._read(1000*1000*5,1000*1000*1)
        print(block)
```

What happens next depends on the server. As a result of the HTTP GET, the server will now send data over the WiFi link and the device will send this over the serial connection as soon as it gets it. Notice that this data is not a direct response to a command and so the device prefixes it with:

`+IPD,len:`

The `+IPD` makes it clear to the client that is a packet of data sent from the server. The `len` value gives the number of characters sent after the colon. You could use this to work out when to stop reading data, but for simplicity we scan for "</html>" which is usually, but not always, the final tag at the end of a web page.

In principle what your program should do next is sit in a polling loop looking for `+IPD`. It should then read the digits between the comma and the colon and convert this to an integer. Finally it should then read exactly that number of characters from the serial port:

```
+IPD,1440:HTTP/1.1 200 OK
Accept-Ranges: bytes
Age: 223181
Cache-Control: max-age=604800
Content-Type: text/html; charset=UTF-8
Date: Sun, 11 Apr 2021 06:40:22 GMT
Etag: "3147526947+gzip"
Expires: Sun, 18 Apr 2021 06:40:22 GMT
Last-Modified: Thu, 17 Oct 2019 07:18:26 GMT
Server: ECS (dcb/7F83)
Vary: Accept-Encoding
X-Cache: HIT
Content-Length: 1256
+IPD,127:<html><head><title>Temperature</title></head>
<body><p>{"humidity":0,"airtemperature":0}
       </p></body></html>CLOSED
```

You can see that there are how many "packets" of data – 1440 characters and 127 characters. In principle you could process the `+IPD` characters as they come in and work out how many characters to read. However, you still wouldn't know how many packets to expect.

The complete function and a more general main program is:

```
 def getWebPage(self,URL,page):
        command= 'AT+CIPSTART="TCP",'+'"'+URL+'"'+",80\r\n"
        self.uart.write(command)
        block=self._read(1000*1000*2,1000*1000)
        if block.find("OK")<0:
            return -1
        command='GET ' + page + ' HTTP/1.1\r\nHost:'+URL+'\r\n\r\n'
        command2="AT+CIPSEND="+str(len(command))+ "\r\n"
        self.uart.write(command2)
        block=self._read(1000*1000*8,1000*1000*1)
        if block.find(">")<0:
            return -1
        self.uart.write(command)
        block=self._read(1000*1000*5,1000*1000*1)
        return block
wifi=WiFi(1,4,5)
response = wifi.mode(1)
response = wifi.connect("ssid","password")
print(wifi.getWebPage("example.com", "/index.html"))
```

This program makes use of some of the functions given earlier.

A Web Server

The most common use for an internet connection on a small device like the Pico is to allow another device to request data. It is fairly easy to create a web server running on the ESP8266, but don't expect Apache or anything advanced. All you can reasonably do is accept a connection and send a web page or two back to the client.

The key differences between client and server mode is that in server mode the device is constantly "listening" for clients to make TCP connections on the port. When the device receives a connection it reads all of the data the client sends and passes it on via the serial port to the Pico. This means that in server mode the Pico has to be constantly on the lookout for new data from the ESP8266. You can do this using an interrupt or a second thread, but for simplicity this example uses a polling loop.

There is another difference between client mode and server mode - there can be multiple TCP connections from as many clients as try to connect. The solution to this problem is that the ESP8266 assigns each TCP socket connection an id number and this is what you need to use to make sure you send the data to the right place.

Let's see how it all works. Assuming we are already connected to WiFi and have an IP address, we can set up a server quite easily. First we need to use the command `CIPMUX=1` to set the device into multiple connection mode. You cannot start a server if `CIPMUX=0`, the default single connection mode. Once multiple connections are allowed you can create a server using `CIPSERVER=1,`*`port`*.

In this example we are using port 80. the standard HTTP port, but you can change this to anything you want:

```
def startServer(self):
    self.uart.write("AT+CIPMUX=1\r\n")
    response=self._read(1000*1000*5,1000*1000*10)
    print("mux set",response)
    if response.find("OK")<0:
        return -1
    self.uart.write("AT+CIPSERVER=1,80\r\n")
    response=self._read(1000*1000*5,1000)
    if response.find("OK")<0:
        return -1
    print("ready for clients",response)
```

If you run just this part of the program you will see the response:

```
mux set AT+CIPMUX=1
OK

ready for clients AT+CIPSERVER=1,80
no change
OK
```

Now we just have to wait for a client to make a connection and send some data. This is done simply by reading the serial input and checking for "+IPD" in an infinite polling loop:

```
while True:
    block=self._read(1000*1000*5,100)
    if block==None:
        continue
    if block.find("+IPD")<0:
        continue
```

If we don't get a block containing +IPD we simply move on to the next iteration, i.e. there is nothing to do.

If we have received a block containing "+IPD", the body of the loop processes it. Note that all of the following instructions are inside the loop and are repeated for each request for the web page. We need to remove the IPD information. The received data has the format:

```
+IPD,id, rest of data
```

We need to extract the ID, as a string, so we can use it to communicate with the client. This is just some standard string handling, but it is still messy:

```
n = block.find("+IPD,")+5
m = block.find(",",n)
id = block[n:m]
```

The algorithm is:

1. Find "+IPD," and set n to the position of the next character
2. Find the next comma and set m to its position
3. Extract the characters from position n to just before position m

Now we have the ID we can communicate with the client, but first we need something to send. As with all HTTP transactions, we have to send some headers and then some data. There are a lot of possible headers you could send, but a reasonable minimum that works with most browsers is:

```
data="HTTP/1.0 200 OK\r\nServer: Pico\r\nContent-type: text/html\r\
n\r\n<html><head><title>Temperature</title></head><body><p>{\"humidi
ty\":81%,\"airtemperature\":23.5C}</p></body></html>\r\n"
```

Note that this is a single string with no spaces between the elements that have been split over different lines for presentation. Of course, in a real application, the HTML part of the data would be generated by the program or read from a file. You can include timestamps and lots of other useful information, but this is simple and it works. Remember to include a blank line at the end of the headers. This is vital as the browser will ignore everything sent to it if you don't.

Now we want to send the data to the client. This is just a matter of using the `CIPSEND` command again, only this time with `id` specified as the first parameter:

```
CIPSEND=id,data length
```

and we check for the response ">" before sending the data. Notice that we don't have to open a TCP socket as we did in the case of acting as a client. The TCP socket has already been opened by the client connecting to the server and when the transaction is complete we can close it:

```
        self.uart.write(command)
        response=self._read(1000*1000*5,1000*1000*1)
        if response.find(">")<0:
            return -1
```

Now, at last, we can send the data to the client:

```
        self.uart.write(data)
        response=self._read(1000*1000*5,1000*1000*1)
        if response.find("OK")<0:
            return -1
```

and wait for it to complete.

Finally we close the connection and complete the loop to wait for another connection:

```
        command="AT+CIPCLOSE="+id+"\r\n"
        self.uart.write(command)
        response=self._read(1000*1000*5,1000*1000*1)
        if response.find("OK")<0:
            return -1
```

Try it out with a main program something like:

```
wifi=WiFi(1,8,9)
response = wifi.mode(1)
print("mode",response)
wifi.connect("SSID","password")
print(wifi.startServer())
```

You should now be able to connect to the IP address that is displayed and retrieve the web page that displays:

```
{"humidity":81%,"airtemperature":23.5C}
```

There are a lot of WiFi commands that haven't been covered here, but now that you have seen examples of most of the basic types and encountered the typical problems that occur, you should be able to implement any that you need.

There is still going to be the occasional unexplained crash and in this case the best solution is to use the soft reset command introduced earlier. It is also worth mentioning that the server program as presented does not actually handle multiple simultaneous connections. It has to finish dealing with one connection before it can deal with a second. To make it deal with multiple connections would mean checking for another connection while processing the first.

Complete Listing Of Web Server

The following listing also includes all of the functions presented in the chapter, even if they aren't used.

```
from utime import  ticks_us, sleep_ms
from machine import UART, Pin

class WiFi:
    def __init__(self,uartNo,tx,rx):
        self.uart = UART(uartNo, baudrate=115200, bits=8, stop=1,
                   parity=None, rx=Pin(rx), tx=Pin(tx))

    def _read(self,timeout, interchar):
        rxData = bytearray(2000)
        t = ticks_us()
        while True:
            if self.uart.any() > 0:
                break
            if ticks_us()-t > timeout:
                return None
        while True:
            while self.uart.any() > 0:
                rxData += self.uart.read(1)
                t = ticks_us()
            if ticks_us()-t > interchar:
                break
        return rxData.decode("utf-8")
```

```
def AT(self):
    self.uart.write(b"AT\r\n")
    return self._read(10000, 100)

def getVersion(self):
    self.uart.write("AT+GMR\r\n")
    return self._read(10000, 100)

def reset(self):
    self.uart.write("AT+RST\r\n")
    print(self._read(10000, 10000))
    return self._read(10000, 10000)

def setUART(self,baud):
    self.uart.write("AT+UART_CUR="+str(baud)+",8,1,0,0\r\n")
    return self._read(10000, 100)

def mode(self, mode):
    command="AT+CWMODE_CUR="+str(mode)+"\r\n"
    self.uart.write(command)
    return self._read(10000, 100)

def scan(self):
    self.uart.write("AT+CWLAP\r\n")
    return self._read(100000, 1000*1000*20)

def scandisplay(self):
    self.uart.write("AT+CWLAP\r\n")
    block=self._read(10000, 100)
    print(block)
    for i in range(20):
        block=self._read(1000*1000*20, 100)
        if block==None:
            break
        if block.find("OK")>=0:
            break
        print(block)

def connect(self,ssid,password):
    command="AT+CWJAP_CUR=" +'"'+ ssid+'"' +","
                                +'"'+password+'"'+ "\r\n"
    self.uart.write(command)
    for i in range(20):
        block=self._read(1000*1000*20, 100)
        if block==None:
            break
        if block.find("OK")>=0:
            break
    print("connected")
```

```
def getIP(self):
    self.uart.write("AT+CIFSR\r\n")
    return self._read(100000, 100)

def getWebPage(self,URL,page):
    command= 'AT+CIPSTART="TCP",'+'"'+URL+'"'+",80\r\n"
    self.uart.write(command)
    block=self._read(1000*1000*2,1000*1000)
    if block.find("OK")<0:
        return -1
    command='GET ' + page + ' HTTP/1.1\r\nHost:'+URL+'\r\n\r\n'
    command2="AT+CIPSEND="+str(len(command))+ "\r\n"
    self.uart.write(command2)
    block=self._read(1000*1000*8,1000*1000*1)
    if block.find(">")<0:
        return -1
    self.uart.write(command)
    block=self._read(1000*1000*5,1000*1000*1)
    return block
```

```
    def startServer(self):
        self.uart.write("AT+CIPMUX=1\r\n")
        response=self._read(1000*1000*5,1000)
        print("mux set",response)
        if response.find("OK")<0:
            return -1
        self.uart.write("AT+CIPSERVER=1,80\r\n")
        response=self._read(1000*1000*5,1000)
        if response.find("OK")<0:
            return -1
        print("ready for clients",response)

        while True:
            block=self._read(1000*1000*5,100)
            if block==None:
                continue
            if block.find("+IPD")<0:
                continue

            n=block.find("+IPD,")+5
            m=block.find(",",n)
            id=block[n:m]

            data="HTTP/1.0 200 OK\r\nServer: Pico\r\nContent-type:
text/html\r\n\r\n<html><head><title>Temperature</title></head><body>
<p>{\"humidity\":81%,\"airtemperature\":23.5C}</p></body></html>\r\
n"
            command="AT+CIPSEND="+id+","+str(len(data))+"\r\n"
            self.uart.write(command)
            response=self._read(1000*1000*5,1000*1000*1)
            if response.find(">")<0:
                return -1
            self.uart.write(data)
            response=self._read(1000*1000*5,1000*1000*1)
            if response.find("OK")<0:
                return -1
            command="AT+CIPCLOSE="+id+"\r\n"
            self.uart.write(command)
            response=self._read(1000*1000*5,1000*1000*1)
            if response.find("OK")<0:
                return -1


wifi=WiFi(1,4,5)
print(wifi.AT())
response = wifi.mode(1)
print("mode",response)
wifi.connect("SSID","password")
print(wifi.startServer())
```

Summary

- Adding WiFi to the Pico is fairly easy using the low-cost ESP8266 ESP-01, which connects via the serial port and makes use of AT style commands to control the device as if it was a WiFi modem.
- In addition to an ESP8266, you also need a power supply capable of running it. The Pico seems able to do the job from its 3.3 output.
- You can use AT commands to set the device into client mode and connect to a WiFi network.
- While it is possible to use ad-hoc protocols, there are advantages in using TCP, HTTP and HTML so that other devices can work with the Pico.
- The Pico can use client mode to download data from web servers. It can also emulate a server to deliver data to any web browser or HTML-using client.

Chapter 18
Direct To The Hardware

MicroPython provides classes and methods to let you access most of the major hardware features of the Pico. They are very simple wrappers around the basic mechanism of working with the hardware – memory-mapped registers. Unfortunately at the time of writing there are many hardware features which are simply not exposed via MicroPython. In most cases it is possible to extend what you access using lower-level interactions with the hardware – still staying in MicroPython, but writing and reading the low-level register based hardware.

The obvious reason for knowing how to use memory-mapped registers is that if MicroPython doesn't provide a function that does what you want, create it! Perhaps a better reason is just to know how things work. In this chapter we take a look at how the Pico presents its hardware for you to use and how to access it via basic software.

Registers

Some processors have special ways of connecting devices, but the Pico's processor uses the more common memory-mapping approach. In this, each external device is represented by a set of memory locations or "registers" that control it. Each bit in the register controls some aspect of the way the device behaves. Groups of bits also can be interpreted as short integers which set operating values or modes.

How do you access a register? MicroPython provides a number of ways of doing this but the simplest is to make use of the `mem` functions in the machine module:

`machine.mem32[address]`	Returns or sets a 32-bit value at the address
`machine.mem16[address]`	Returns or sets a 16-bit value at the address
`machine.mem8[address]`	Returns or sets an 8-bit value at the address

The only difficult part is in working out the address you need to use and the value that sets or resets the bits you need to modify.

For example, if you look in the documentation you will find that the GPIO registers start at address `0x40014000`. The registers are defined by their offset from this starting address. So for example, the table of GPIO registers is:

Offset	Register Name	Description
`0x000`	GPIO0_STATUS	GPIO status
`0x004`	GPIO0_CTRL	GPIO control including function select and overrides
`0X008`	GPIO1_STATUS	GPIO status
`0x00c`	GPIO1_CTRL	GPIO control including function select and overrides
... and so on down to		
`0x0ec`	GPIO29_CTRL	GPIO control including function select and overrides

You can see that there are two registers for each GPIO line from GP0 to GP29, one control register and one status register.

Each register has the same format for each GPIO line. For example, the status register is:

Bits	Name	Description	Type	Reset
31:27	Reserved		-	-
26	IRQTOPROC	Interrupt to processors, after override applied	RO	`0x0`
25	Reserved		-	-
24	IRQFROMPAD	Interrupt from pad, before override applied	RO	`0x0`
23:20	Reserved		-	-
19	INTOPERI	Input signal to peripheral, after override applied	RO	`0x0`
18	Reserved		-	-
17	INFROMPAD	Input signal from pad, before override applied	RO	`0x0`
16:14	Reserved		-	-
13	OETOPAD	Output enable to pad, after override applied	RO	`0x0`
12	OEFROMPERI	Output enable from selected peripheral, before override applied	RO	`0x0`
11:10	Reserved		-	-
9	OUTTOPAD	Output signal to pad after override applied	RO	`0x0`
8	OUTFROMPERI	Output signal from selected peripheral, before override applied	RO	`0x0`
7:0	Reserved		-	-

You can see that many of the 32 bits in the register are not used, but bit 9 is OUTTOPAD which is the final state of the GPIO line after register overrides have been applied. You can read its current value using:

```
from machine import mem32
addrGP0Status= 0x40014000
value=mem32[addrGP0Status]
print(bin(value))
```

This prints the current status of `GP0` in binary. If you want to find the status of `GPn` you need to use address `0x40014000+2n`. Usually addresses are specified as a base address, i.e. where things start, and an offset that has to be added to the base to get the address of a specific device.

This is the general way you work with peripheral devices such as the PWM units or I2C hardware, but the GPIO is special in that it has another set of registers that control it.

Single-Cycle IO Block

At this point you might think that we are ready to access the state of the GPIO lines for general input and output. This isn't quite the whole story. To accommodate the fact that the processor has two cores, and to make access faster to important devices, there is a special connection, the SIO or Single-cycle IO Block, between the cores and, among other things, the GPIO. The SIO connects directly to the two cores and they can perform single-cycle 32-bit reads and writes to any register in the SIO. Notice that the SIO is not connected via the general address bus.

You can see the general structure of the SIO in the diagram below. You can find out about the other devices it connects to from the documentation - our focus is on the GPIO lines.

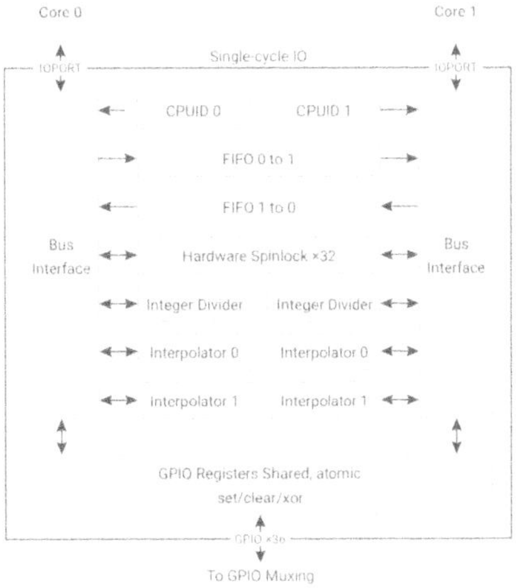

Notice that the GPIO lines are multipurpose and the SIO only has control when they are being used as GPIO lines. In this sense the SIO is just another peripheral that can take control of a GPIO line.

The SIO provides a set of registers that makes using the GPIO much faster and much easier. The basic registers are:

`GPIO_OUT` Sets all GPIO lines to high or low

`GPIO_IN` Reads all GPIO lines

`GPIO_OE` Sets any GPIO line to output driver or high impedance

There are also three registers – SET, CLR and XOR - that make working with GPIO_OUT and GPIO_OE easier. Each of these can be thought of as a mask that sets, clears or XORs bits in the corresponding register.

For example, GPIO_OUT_SET can be used to set just those bits in GPIO_OUT that correspond to the positions that are set high.

The locations of these registers are as offsets from `0xd0000000`:

Offset	Name	Description
0x004	GPIO_IN	GPIO Input value
0x010	GPIO_OUT	GPIO output value
0x014	GPIO_OUT_SET	GPIO output value set
0x018	GPIO_OUT_CLR	GPIO output value clear
0x01c	GPIO_OUT_XOR	GPIO output value XOR
0x020	GPIO_OE	GPIO output enable
0x024	GPIO_OE_SET	GPIO output enable set
0x028	GPIO_OE_CLR	GPIO output enable clear
0x02c	GPIO_OE_XOR	GPIO output enable XOR

Blinky Revisited

Now we can re-write Blinky yet again, but this time using direct access to the SIO GPIO registers.

```
from machine import mem32,Pin
from time import sleep_ms
led=Pin(25,mode=Pin.OUT)
addrSIO = 0xd0000000
while True:
    mem32[addrSIO + 0x014] = 1 << 25
    sleep_ms(500)
    mem32[addrSIO + 0x018] = 1 << 25
    sleep_ms(500)
```

This program uses the standard MicroPython class to set the GPIO line to SIO control and output. If you think that this is cheating, it is an exercise to set the line correctly using the GPIO control register and the SIO.

This example is a demonstration rather than being useful, but there are some very useful functions we can write using our knowledge of how the GPIO lines are controlled. MicroPython is limited to controlling a single GPIO line at a time, but the hardware can change or read multiple GPIO lines in a single register operation. For example:

```
def gpio_get():
    return mem32[0xd0000000+0x010]
```

Here the get function simply reads the GPIO_OUT register which has a single bit for the output state of each GPIO line. Notice that GPIO lines set to output reflect their last written-to state.

A set function simply writes the mask to the GPIO_OUT_SET register

```
def gpio_set(mask):
        mem32[0xd0000000+0x014] = mask
```

A clear function is just as easy and this just writes to the GPIO_OUT_CLR register:

```
def gpio_clear(mask):
        mem32[0xd0000000+0x18C] = mask
```

You can easily create functions for reset and other logical operations on all of the GPIO lines in one operation, but a single mask value function is usually sufficient:

```
def gpio_set(value,mask):
        mem32[0xd0000000+0x01C] =
(machine.mem32[0xd0000000+0x010])^value & mask
```

This writes to the GPIO_OUT_XOR register, but it writes a combination of a mask and a value. The `mask` gives the GPIO lines that need to be changed and the `value` gives the state they are to be set to. For example, if `mask` is `0111` and `value` is `0100` then `value & mask` is `0100`. If this is XORed with the current state of the lines – e.g. 0101, in this case the result is `0001`, which changes the state of only GP0 to a zero. Thus we have set lines GP2, GP1 and GP0 as specified in the mask to the corresponding bits in the value, i.e. `0100`. Notice that this process sets the lines selected in the mask to either a zero or a one as determined by the bits in value.

As demonstrated in Chapter 4, the `value,mask` function can be used to set GPIO lines simultaneously:

```
from machine import Pin
import machine

def gpio_get():
    return machine.mem32[0xd0000000+0x010]

def gpio_set(value,mask):
     machine.mem32[0xd0000000+0x01C]=
                (machine.mem32[0xd0000000+0x010])^value & mask


pin=Pin(22,Pin.OUT)
pin=Pin(21,Pin.OUT)
value1=1<<22 | 0<<21
value2=0<<22 | 1<<21
mask=1<<22 | 1<<21
while True:
    gpio_set(value1,mask)
    gpio_set(value2,mask)
```

This sets lines GP21 and GP22 to `01` and `10` on each pass through the loop:

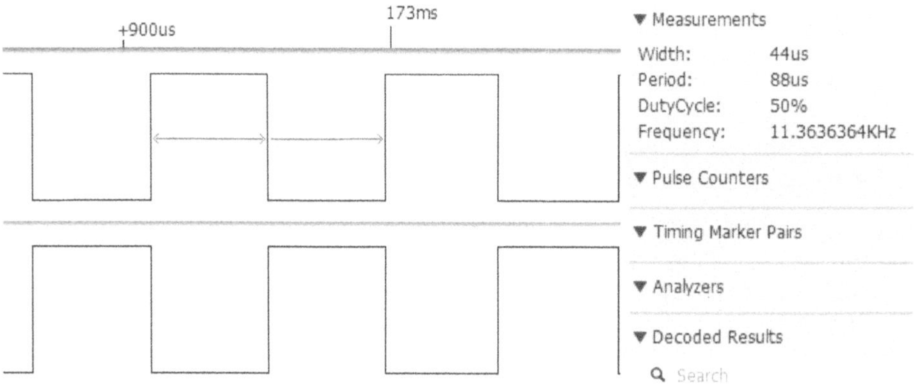

Example I - Events

In Chapter 7 the idea of events was introduced, but MicroPython doesn't provide any access to events. The solution is to add our own function that accesses the GPIO register that records interrupts. This is a register in the GPIO set of registers rather than in the SIO as GPIO interrupts aren't specific to what is controlling the GPIO line.

The base address of the GPIO registers is `0x40014000.` After the end of the set of status and control registers, starting at offset `0x0f0`, there is a block of four interrupt registers that record the status of the GPIO lines. Each group of four bits gives the status of the various level and edge events:

Bits	Name	Type	Reset
31	GPIO7_EDGE_HIGH	WC	0x0
30	GPIO7_EDGE_LOW	WC	0x0
29	GPIO7_LEVEL_HIGH	RO	0x0
28	GPIO7_LEVEL_LOW	RO	0x0
27	GPIO6_EDGE_HIGH	WC	0x0
26	GPIO6_EDGE_LOW	WC	0x0
25	GPIO6_LEVEL_HIGH	RO	0x0
24	GPIO6_LEVEL_LOW	RO	0x0
	. . .		
8	GPIO2_LEVEL_LOW	RW	0x0
7	GPIO1_EDGE_HIGH	RW	0x0
6	GPIO1_EDGE_LOW	RW	0x0
5	GPIO1_LEVEL_HIGH	RW	0x0
4	GPIO1_LEVEL_LOW	RW	0x0
3	GPIO0_EDGE_HIGH	RW	0x0
2	GPIO0_EDGE_LOW	RW	0x0
1	GPIO0_LEVEL_HIGH	RW	0x0
0	GPIO0_LEVEL_LOW	RW	0x0

The format of the raw interrupt registers is more complicated than the previous one bit to one GPIO line arrangement we have encountered before. In this case there are four bits per GPIO line and they record different interrupt types. The first four bits of the first register record interrupts on GP0:

3	GPIO0_EDGE_HIGH	WC	0x0
2	GPIO0_EDGE_LOW	WC	0x0
1	GPIO0_LEVEL_HIGH	RO	0x0
0	GPIO0_LEVEL_LOW	RO	0x0

This pattern is repeated for each of the GPIO lines and, when all of the bits in the first register have been used, the pattern continues in the second register with GP8 and so on. Each register records the event data for eight GPIO lines. Notice that each of the bits is set if the event that would cause the interrupt occurs – the interrupt itself only occurs if it is enabled. What this means is that the level bits track the current level of the GPIO line and the edge bits are set if an edge of that type has occurred. The WC in the third column indicates that the bit is cleared if you write to it and this is how the event is cleared.

The new problem here is that we have to work out which register is concerned with which GPIO line – each register looks after eight GPIO lines – and which group of four bits in the register gives the events for that GPIO line.

The following function accepts the GPIO number and works out which register and group of bits it corresponds to:

```
def gpio_get_events(pinNo):
    mask = 0xF << 4 * (pinNo % 8)
    intrAddr = 0x40014000 + 0x0f0 + (pinNo // 8)*4
    return (machine.mem32[intrAddr] & mask) >> (4 * (pinNo % 8))
```

The calculation for the address of the register needed is just the base address plus the offset of the first interrupt register and then `+(pinNo//8)*4`. As each register deals with a group of 8, `pinNo//8` (integer division) gives the number of the register needed and `*4` converts this to a byte address, i.e. each register is four bytes. The mask is constructed using a similar technique. The first GPIO line uses four bits starting at 0, the second uses four bits starting at 4 the third at 8 and so on, i.e. `(4*(pinNo % 8)`, and this is used to create a mask.

Once you have the basic way of accessing the bits you need, you can reuse it to create a clear events function:

```
def gpio_clear_events(pinNo, events):
    intrAddr = 0x40014000 + 0x0f0 + (pinNo // 8)*4
    machine.mem32[intrAddr] = events << (4 * (pinNo % 8))
```

To clear the raw interrupt bit you simply have to write a zero to it.

Example II PAD - Pull, Drive and Schmitt

Each GPIO line has an identical input output stage, called a PAD, which is the connection to the outside world, no matter what mode the GPIO line is being used in. This is fundamental to the workings of the GPIO line and you might be wondering why it is being introduced so late? The answer is that MicroPython doesn't fully support it and the aspects of the PAD that it does support, Pull Up and Pull Down, are simple enough. If you are interested in the finer details then you are going to have to implement functions that work with them.

The structure of the PAD can be seen below:

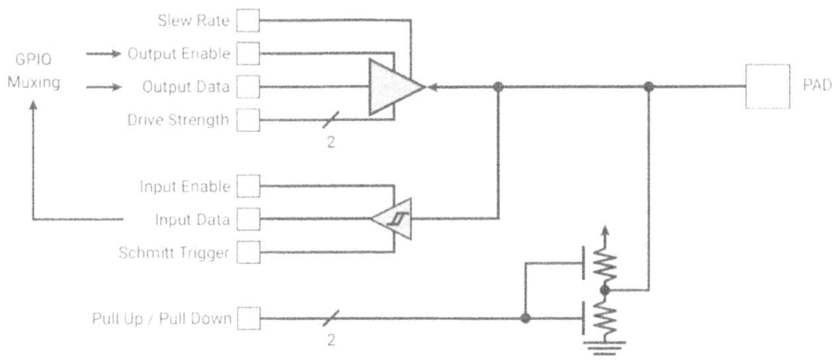

You can see that under program control you can set the pull up/pull down configuration and enable the input/output. The input also has a Schmitt trigger that can be enabled to clean up noisy inputs. The output can be customized by slew rate, how fast it changes and drive strength.

Before moving on to the software, it is worth explaining the basic ideas of the options. The Schmitt trigger adds hysteresis to the input line. This means that before the state changes from high to low it has to cross a threshold, but to change back to a low state it has to cross a lower threshold. This acts like a limited debounce mechanism in that it stops the line from going low because the input drops a little after going high. The Pico's Schmitt trigger uses thresholds of a 0.2V difference if the processor supply voltage is 2.5V to 3.3V and 0.18V if the voltage is 1.8V. What this means is that at 3.3V the input has to be greater than 2V to be a one, but after this the voltage has to fall to 1.8V before it changes back to a zero. For a zero, the thresholds are 0.8V to change a zero and 2V to change back to a one.

The output drive strength isn't to do with how much current the GPIO line can source, it is about the voltage output at different currents. It is the effective output resistance. Each time the drive current is increased by 2mA another transistor is used in the drive, so lowering the output resistance. This has the effect of increasing or decreasing the voltage at the pin. For example, if you set the drive to 1 then, if you want to keep the output voltage at or above 2.62V, i.e. a logic 1, you can't draw more than 2mA. Put more simply, if you want the output to maintain voltages that are within the threshold for a logic 1 or 0 then you can only draw 2, 4, 8 or 12mA from the GPIO line depending on the setting of the drive strength. If you draw more current, this is fine, but the voltage will fall below the standards for 3.3V logic.

The final option is to change the slew rate. There is no information on this in the datasheet, but slowing the rate of change of the GPIO line can be useful when driving long lines or reactive loads. How useful this is in practice is a matter of experimentation.

If you want to work with the PAD directly then you need to know the location and format of the PAD control registers. These start at `0x4001c000` and the first, controlling the GP0 PAD is at offset `0x04` and in general the register for GPn PAD is at offset `4(n+1)`. The format of the PAD register is:

Bits	Name	Description	Type	Reset
31:8	Reserved	-	-	-
7	OD	Output disable. Has priority over output enable from peripherals	RW	0x0
6	IE	Input enable	RW	0x1
5:4	DRIVE	Drive strength. 0x0 → 2mA 0x1 → 4mA 0x2 → 8mA 0x3 → 12mA	RW	0x1
3	PUE	Pull up enable	RW	0x0
2	PDE	Pull down enable	RW	0x1
1	SCHMITT	Enable Schmitt trigger	RW	0x1
0	SLEWFAST	Slew rate control. 1 = Fast, 0 = Slow	RW	0x0

Using this information it is fairly easy to write functions to set each of the characteristics of the PAD. For example, to read the slew rate:

```
def pad_get_slew(gpio):
    return mem32[0x4001c000+(gpio+1)*4] & 0x01
```

To set the skew rate:

```
def pad_set_slew(gpio,value):
    if value:
        mem32[0x4001c000+(gpio+1)*4]=
                        mem32[0x4001c000+(gpio+1)*4] | 0x1
    else:
       mem32[0x4001c000+(gpio+1)*4]=
                        mem32[0x4001c000+(gpio+1)*4] & 0xFFFFFFFE
```

Getting and setting the Schmitt trigger is just as easy, but now we set or clear the second bit not the first:

```
def pad_set_schmitt(gpio,value):
    if value:
        mem32[0x4001c000+(gpio+1)*4]=
                        mem32[0x4001c000+(gpio+1)*4] | 0x2
    else:
        mem32[0x4001c000+(gpio+1)*4]=
                        mem32[0x4001c000+(gpio+1)*4] & 0xFFFFFFFD
```

Setting the drive is slightly more complicated as it is a three-bit value:

```
def pad_get_drive(gpio):
    return (mem32[0x4001c000+(gpio+1)*4] & 0xE0) >> 5


def pad_set_drive(gpio, value):
    mem32[0x4001c000+(gpio+1)*4]=
                mem32[0x4001c000+(gpio+1)*4] ^ (0xE0 & (value<< 5))
```

The set function uses the same logic as the value mask function given earlier to only change the bits we are interesting in changing.

Of course we, don't need to deal with the pull-ups or pull-downs as there are methods that do the job.

Digging Deeper

There is so much more to explore, but now we have covered the details that make things difficult to get started. From here you can read the datasheet to find out how the registers control things and implement MicroPython functions to extend what you can do. The biggest difficult is finding the register that contains the bits that reflect the status or control whatever it is you are interested in. Once you have found this out, the only remaining difficulty is in working out how to set or clear the bits you need to work with without changing other bits. It also has to be said that hardware documentation at this level is often incomplete due to assumptions the writer makes about what you should already know. In such a circumstance your best approach is the experimental method. Work out the simplest program you can think of to verify that you understand what the hardware does – and if you are wrong always check the addresses and bits you are changing before concluding that things work differently to the manual.

Summary

- All of the peripherals, including the GPIO lines, are controlled by registers – special memory locations that you write and read to configure and use the hardware.
- Exactly where the registers are positioned in address space is given in the documentation as a base address used for all of the similar registers and an offset that has to be added to the base to get the address of a particular register.
- The SIO block provides a more convenient way to access the GPIO lines and it has a different set of addresses and registers to the GPIO lines.
- With knowledge of how things work, you can add functions that are missing from MicroPython such as events and PAD control.
- Each GPIO line connects to the outside world via a PAD which has a number of configurable elements such as pull-up, slew rate and so on.

Index

1-Wire....228
2N2222....15, 47, 49, 109
2N7000....51
775 motor....107
8n1....260

ADC....27, 153
address....176
aliasing....64
AM2302....209
Analog to Digital Converter....27
any()....265
armature....106
Array.array....83
AT command....273
AT commands....271
autopull....197
autopush....202, 224

base....46
baud....259, 260
beta....46
bidirectional....112
bipolar....125
bit banging....185
BJT....46, 47, 50, 57, 109
Blinky....21, 30, 31, 194, 296
BootSel button....20
breakout board....174
brushed motor....106
brushless DC motor....122
busy wait....37
button....54, 60, 68, 79
bytearray....170

C....35
checksum....181, 236, 237
clock divider....190
clock speed....167
clock stretching....173
collector....46
command....176
commutator....106
CoreMeltdown....68
CPHA....139
CPOL....139

CRC....181, 236, 237
current-limiting resistor....31, 44, 48
cyclic redundancy checksum....181, 236, 237

D-to-A....97
Darlington pair....108
DC motor....103
debounce....56, 62
decorator....189
deinit....92, 168
DHT11....209
DHT22....209, 217
dip relay....49
disable_irq....87
drain....50
Drain-Source On-State Resistance....50
DS18B20....228, 239, 250, 256
Dupont wires....15
duty cycle....90, 95, 97
Duty Cycle Resolution....96
duty_ns....91
duty_u16....91

EEPROM....256
Electronic Speed Controller....123
emitter....46
ESC....123
ESP-01....271
ESP-07....272
ESP8266....271
Espressif Systems....271
event....76
event queue....75
fast mode....167

FET Field Effect Transistor....47, 50, 57
FIFO....75, 218, 224
finite state machine....67
fixed time delay....37
floating....54
forward voltage drop....44
freq....91
FSM....67, 70, 72
FSM Button....68
FSM Hold Button....69
FSM Ring Counter....70
FTP....25

gate....50
gearbox....124
GPIO....27, 33, 69
GPIO registers....294
gpio_get....40
gpio_set....40

H-bridge....109, 113, 115, 130
half bridge....117
hard....82, 85
hfe....48
hold....63
Hold master....177
how fast....34
HTTP....282
HTU21D....174
humidity....180, 215

I2C....53, 165, 168
iButton....228
IN....60, 201
infinite loops....66
init....54, 60, 141
input....59
Input Shift Register....201
inrunner,....122
integration....56
inter-character timeout....267
internal pull-up....61
interrupt....76, 82, 85, 300
interrupt service routine....76
inverting buffer....57
IRFZ44NPBF....52, 109
irq....82
ISR....76, 82, 201, 218, 224

jitter....96

L298 Dual H-bridge....114
lack of speed....19
LED....15, 20, 28, 30, 31, 44, 45, 46, 47, 61, 100, 102, 103
logic analyzer....15, 31, 34
loopback....143
low-pass filter....97

mask....40, 297
MAX14870....117
MCP3008....156
Mealy machine....67

mem....293
micro-steps....126
MicroPython....13, 14, 19
middle C....99
MISO....137
MOSFET....50, 52, 109, 112, 113, 127
MOSI....137
motor....105
multibyte....143, 169
multimeter....15

N-channel....50
No Hold master....177, 180
non-linear....44
NPN....46

Ohm's law....43, 44
ohmic....44
open collector....209
open collector bus....166
OPEN_DRAIN....54
oscilloscope....34
OSR....195
OUT....196
output drive strength....302
output resistance....302
outrunner....122
OUTTOPAD....295

P-channel....50
parasitic....228
parity bits....260
password....281
Permanent Magnet....124
phase....141
phase-correct....93
phased pulses....38
Pico....11
Pico-Go....24
Pin object....33
PIO....11, 185, 217, 247
PIO clock....190
PM stepper....124
PNP....47
polarity....94, 141
polling....180, 265
polling is wasteful....66
polling loop....59, 63, 66, 68, 72, 77, 173
potential divider....56

potentiometer 118
power consumption 43
press 63
pull-down 52, 53
pull-up 52, 53, 55, 166, 230
Pulse Width Modulation 89
push 202
push-pull 52
PWM 28, 89, 95, 100, 103, 108, 115, 119
PWM applications 97
PWM Pin 91
Pylance 23
Python 3 19

race conditions 86
read 142, 263
Read ROM 255
read_u16 155
readfrom 171
reading from a register 171
readinto 141, 142, 263
readline 263
register 300
resistors 15
rotor 105
rp2 module 187
rp2.asm_pio 187
RP2040 13
RS232 259
RX FIFO 264

sample-and-hold 157
Schmitt trigger 301
SCL 166
SCLK 137, 138
ScratchPad 255
SDA 166
Search ROM 255
serial bus 165
serial port 259
servo 124
set_init 189
setPosition 119
SETSIDES 199
Si7021 174
SIO 299
sleep 36
sleep_ms 36
sleep_us 36

slew rate....302
Slow Read....173
SN754410....115
SoC....11
soft reset command....278, 288
software interrupt....77
solderless prototype board....15
source....50
speed....19
SPI....158, 159
SPI bus....137, 153
SPI module....140
SPI object....140
spin wait....37, 191
split ring....106
SSID....281
Stack Overflow....16
START....172
starvation....86
state machine....186
StateMachine class....187
stator....105, 122
stepForward....129
stepper motor....124
stepReverse....129
STOP....172
stop bits....260
strong pull-up....228
successive approximation converter....154, 157
surface-mount....174
switch....54, 60
synchronization....95

TCP....282
temperature....179, 215, 243
Thonny....14, 21
ticks_add....38
ticks_diff....38
ticks_ms()....38
ticks_us()....38
time....36
time_pulse_us....65
timeout....267
timer....131
timestamp....82
timing....59
TIP120....108
torque....106
transistor....47

trap........77
tri-state........151
TTY........259
TX FIFO........193, 265, 268

UART........259, 261
UART object........262
uf2 file........20
unidirectional driver........125
unipolar........125
USB connection........20
utime........36

V24........259
Variable Reluctance........124
variable resistor........118
VGS(th)........50, 51
VR........124
VS Code........14, 23

web client........282
web server........285
Weber-Fechner law........100
WiFi........271
wrap........96, 206
write........142, 170
write_readinto........141
writeto........168
writevto........169
writing to a register........170

XOR........238

Other Books by Harry Fairhead

Programming The Raspberry Pi Pico in C

ISBN: 978-1871962680

The C version of this book explains the many reasons for wanting to use C with the Pico, not least of which is the fact that it is much faster. This makes it ideal for serious experimentation and delving into parts of the hardware that are otherwise inaccessible. Using C is the way to get the maximum from the Pico and to really understand how it works.

Raspberry Pi IoT in C, Second Edition

ISBN: 9781871962635

In Raspberry Pi IoT in C you will find a practical approach to understanding electronic circuits and datasheets and translating this to code, specifically using the C programming language. The main reason for choosing C is speed, a crucial factor when you are writing programs to communicate with the outside world. If you are familiar with another programming language, C shouldn't be hard to pick up. This second edition has been brought up-to-date and focuses mainly on the Pi 4 and the Pi Zero. There is new material on the recently introduced GPIO character driver and using the Pi 4's additional ports and scheduling.

The main idea in this book is to not simply install a driver, but to work directly with the hardware using the Raspberry Pi's GPIO (General Purpose Input Output) to connect with off-the-shelf sensors. It explains how to use its standard output with custom protocols, including an in-depth exposition of the 1-Wire bus. You will also discover how to put the Internet into the IoT using sockets.

Although NetBeans is used to develop programs, VS Code is now considered an alternative remote development environment and all the book's code, which is available for download, has been tested with VS Code.

Raspberry Pi IoT in Python With GPIO Zero

ISBN:9781871962666

Python is an excellent language to learn about the IoT or physical computing. It might not be as fast as C, but it is much easier to use for complex data processing. The GPIO Zero library is the official way to use Python with the GPIO and other devices. This book looks at how to use it to interface the Raspberry Pi 4 and Raspberry Pi Zero to IoT devices and at how it works so that you can extend it to custom devices. Studying GPIO Zero is also a great way to improve your Python.

Raspberry Pi IoT in C With Linux Drivers

ISBN:9781871962642

There are Linux drivers for many off-the-shelf IoT devices and they provide a very easy-to-use, high-level way of working. The big problem is that there is very little documentation to help you get started. This book explains the principles so that you can tackle new devices and provides examples of using external hardware via standard Linux drivers with the Raspberry Pi 4 and Raspberry Pi Zero in the C language, which provides optimal performance.

Raspberry Pi IoT in Python With Linux Drivers

ISBN:9781871962659

If you opt to use Linux drivers to connect to external devices then Python becomes a good choice, as speed of execution is no longer a big issue. This book explains how to use Python to connect to and control external devices with the Raspberry Pi 4 and Raspberry Pi Zero using the standard Linux drivers.

Applying C For The IoT With Linux

ISBN: 978-1871962611

If you are using C to write low-level code using small Single Board Computers (SBCs) that run Linux, or if you do any coding in C that interacts with the hardware, this book brings together low-level, hardware-oriented and often hardware-specific information.

It starts by looking at how programs work with user-mode Linux. When working with hardware, arithmetic cannot be ignored, so separate chapters are devoted to integer, fixed-point and floating-point arithmetic. It goes on to the pseudo file system, memory-mapped files and sockets as a general-purpose way of communicating over networks and similar infrastructure. It continues by looking at multitasking, locking, using mutex and condition variables, and scheduling. Later chapters cover managing cores, and C11's atomics, memory models and barriers and it rounds out with a short look at how to mix assembler with C.

Fundamental C: Getting Closer To The Machine

ISBN: 978-1871962604

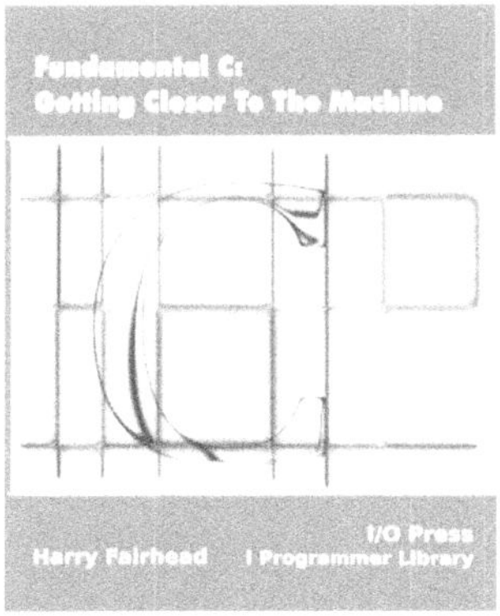

At an introductory level, this book explores C from the point of view of the low-level programmer and keeps close to the hardware. It covers addresses, pointers, and how things are represented using binary and emphasizes the important idea is that everything is a bit pattern and what it means can change.

For beginners, the book covers installing an IDE and GCC before writing a Hello World program and then presents the fundamental building blocks of any program - variables, assignment and expressions, flow of control using conditionals and loops.

When programming in C you need to think about the way data is represented, and this book emphasizes the idea of modifying how a bit pattern is treated using type punning and unions and tackles the topic of undefined behavior, which is ignored in many books on C. A particular feature of the book is the way C code is illustrated by the assembly language it generates. This helps you understand why C is the way it is. And the way it was always intended to be written - close to the metal.

Micro:bit IoT In C, Second Edition

ISBN: 978-1871962673

The new V2 version of the micro:bit is fully covered in this fully updated second edition which now uses VS Code for offline development.

It covers how to get started the easy way with two downloadable templates. Next it looks how to control the micro:bit's I/O lines and explores using the GPIO. For speed we need to work directly with the raw hardware and also master memory-mapping and pulse width modulation. Sensors are connected using first the I2C bus, then by implementing a custom protocol for a 1-Wire bus, and eventually adding eight channels of 12-bit ADC. The micro:bit lacks WiFi connectivity, but can become an Internet server via its serial port. The book concludes with a look at the micro:bit's LED display and how it can be used in C for the classic game of Commando Jump, and a new chapter on the micro:bit's radio and the V2's sound capabilities.

Programmer's Python: Everything is an Object:
Something Completely Different by Mike James

ISBN: 978-1871962581

This book sets out to explain the deeper logic in the approach that Python 3 takes to classes and objects. The subject is roughly speaking everything to do with the way Python implements objects. That is, in order of sophistication, metaclass; class; object; attribute; and all of the other facilities such as functions, methods and the many "magic methods" that Python uses to make it all work. This is a fairly advanced book in the sense that you are expected to know basic Python. However, it tries to explain the ideas using the simplest examples possible. As long as you can write a Python program, and you have an idea what object-oriented programming is about, it should all be understandable and, as important, usable.

Made in United States
Cleveland, OH
05 November 2026

47670542R00172